Sustainable Crop Production - Recent Advances

Edited by Vijay Singh Meena, Mahipal Choudhary, Ram Prakash Yadav and Sunita Kumari Meena

Published in London, United Kingdom

Sustainable Crop Production - Recent Advances
http://dx.doi.org/10.5772/intechopen.98155
Edited by Vijay Singh Meena, Mahipal Choudhary, Ram Prakash Yadav and Sunita Kumari Meena

Contributors
Addisu Fufa Ebbisa, Melkamu Jate, Joachim Lammel, Sheikh Mansoor, Nafeesa Farooq Khan, Aatifa Rasool, Sana Saleem, Tawseef Rehman Baba, Sheikh Marifatul Haq, Sheikh Aafreen Rehman, Simona Mariana Popescu, Charles Oluwaseun Adetunji, Nura Jafar Shanono, Nuraddeen Mukhtar Nasidi, Habibu Ismail, Nura Yahaya Usman, Mu'azu Dantala Zakari, Shehu Idris Umar, Sunusi Abubakar Amin, Priyanka Saha, Anurag Bera, Anamika Barman, Robert M. Merton Stwalley III, Michael M. Boland, Young U. Choi, Daniel G. Foley, Matthew S. Gobel, Nathan C. Sprague, Santiago Guevara-Ocana, Yury A. A Kuleshov, Garima Awasthi, Kumud Kant Awasthi, Mahipal Singh Sankhla, Pritam P. Pandit, Vanisree C. R., Seema Manwani, Chandra Sekhar Yadav, Vibha Jiman, Shashank Patel

First published in London, United Kingdom, 2022 by IntechOpen
IntechOpen is the global imprint of INTECHOPEN LIMITED, registered in England and Wales, registration number: 11086078, 5 Princes Gate Court, London, SW7 2QJ, United Kingdom

British Library Cataloguing-in-Publication Data
A catalogue record for this book is available from the British Library

Additional hard and PDF copies can be obtained from orders@intechopen.com

Sustainable Crop Production - Recent Advances
Edited by Vijay Singh Meena, Mahipal Choudhary, Ram Prakash Yadav and Sunita Kumari Meena
p. cm.
Print ISBN 978-1-80355-696-3
Online ISBN 978-1-80355-697-0
eBook (PDF) ISBN 978-1-80355-698-7

For EU product safety concerns:
IN TECH d.o.o., Prolaz Marije Krucifikse Kozulić 3, 51000 Rijeka, Croatia,
info@intechopen.com or visit our website at intechopen.com.

Meet the editors

Vijay Singh Meena has worked in various aspects of soil aggregation, carbon management index, and carbon and nitrogen sequestration potential under climate-resilient agriculture. He identified the carbon management index as the key indicator to measure soil degradation in different agro-ecosystems.

Mahipal Choudhary has worked on the efficient utilization of soil, water, and nutrient resources in different cropping systems under rainfed and irrigated conditions. He has worked in long-term field evaluation of carbon and nitrogen sequestration, yield sustainability, and chemical and soil biological indicators in the pulse-cereal cropping system.

Ram Prakash Yadav has worked on tree-crop interaction, carbon sequestration potential, forage production, and wasteland management under different land-use types including agroforestry and cropping systems. He identified different multipurpose tree species that can be integrated into agroforestry.

Sunita Kumari Meena has worked on the transformation and availability of phosphorus and sulfur in relation to soil organic carbon under long-term nutrient supply options. Currently, she is an Assistant Professor of Soil Science at the Sugarcane Research Institute (SRI), Dr. Rajendra Prasad Central Agricultural University, Pusa-Samastipur (Bihar), India. She served as a research scholar at both Banaras Hindu University, Varanasi, and at ICAR-Indian Agricultural Research Institute, New Delhi.

Contents

Preface

Sustainable food production systems have a major role to play in ensuring global food security, especially for developing nations. These systems are responsible for producing food that is nutritious and affordable to a rapidly growing global population. Climate change is having major adverse impacts on agricultural production systems, and agriculturists must adapt to and mitigate these negative effects. Therefore, sustainable food production systems are essential.

This book includes eight chapters addressing various nutrient, crop, and soil management issues, including recent advances in sustainable food production in the context of the changing climate. Chapters present case studies on long-term field experiments in specific locations with a focus on the state of the art of sustainable agriculture production systems.

This book is useful for undergraduate and graduate students, teachers, and researchers, particularly in the fields of crop science, soil science, and agronomy.

Vijay Singh Meena
CIMMYT – Borlaug Institute for South Asia (BISA),
Pusa, India

Mahipal Choudhary
ICAR-Central Arid Zone Research Institute,
Jodhpur, India

Ram Prakash Yadav
Rani Lakshmi Bai Central Agricultural University,
Jhansi, India

Sunita Kumari Meena
Dr. Rajendra Prasad Central Agricultural University,
Pusa, India

Section 1

Nutrient Management

Chapter 1

Toward the Recent Advances in Nutrient Use Efficiency (NUE): Strategies to Improve Phosphorus Availability to Plants

Addisu Ebbisa

Abstract

Achieving high nutrient use efficiency (NUE) and high crop productivity has become a challenge with increased global demand for food, depletion of natural resources, and deterioration of environmental conditions. Higher NUE by plants could reduce fertilizer input costs, decrease the rate of nutrient losses, and enhance crop yields. Nitrogen and Phosphorus are the most limiting nutrients for crop production in many of the world's agricultural areas, and their efficient use is important for the economic sustainability of cropping systems. Furthermore, the dynamic nature of N and P in soil-plant systems creates a unique and challenging environment for its efficient management. Although numerous fertilizer recommendation methods have been proposed to improve NUE, technologies and innovative management practices are still lacking. Therefore, maximizing crop phosphorus (P) use efficiency (PUE) would be helpful in reducing the use of inorganic phosphorus fertilizers and their escape in the environment for sustainable agriculture. Improvement of PUE in cropping systems can be achieved through two main strategies: optimizing agronomic practice and breeding nutrient efficient crop cultivars that improves P-acquisition and -utilization efficiency. These strategies are needed for future food security and sustainable agriculture. The major revised points are the following: concept of NUE, application of nutrient stewardship, cereal-legume intercropping, regulating soil pH, etc., for enhancing phytoavailability of P and breeding P-efficient crop cultivars that can produce more biomass with lesser P costs and that acquire more P in P-stress condition. These approaches consider economic, social, and environmental dimensions essential to sustainable agricultural systems and afford a suitable context for specific NUE indicators.

Keywords: agronomic strategies, crop productivity, nutrient acquisition, NUE, PUE, sustainable agriculture

1. Introduction

For sustainable food production, it is an absolute requirement that nutrients removed with the harvest of crops are replaced to prevent nutrient depletion and soil degradation. Achievement and maintenance of high nutrient use efficiency (NUE)

together with high crop productivity have become a major challenge in both developed and developing countries with an increasing growing population, depletion of natural resources, and deteriorating environmental conditions. This is occurring at the same time as society becomes ever more concerned about resource management practices and the environment, especially when it comes to nutrient management [1]. Fertilizer nutrients applied that are not taken up by the crop are also vulnerable to losses from leaching, erosion, and denitrification or volatilization in the case of N, or they could be temporarily immobilized in soil organic matter to be released at a later time, all of which impact apparent use efficiency [2].

Improving nutrient use efficiency (NUE) in plants is vital to enhance the yield and quality of crops, reduce nutrient input cost and improve soil, water, and air quality [3]. Higher NUE by plants could reduce fertilizer input costs, decrease the rate of nutrient losses, and enhance crop yields. Improving crop nutrient use efficiency ideally requires an understanding of the whole system, from the macro (agroecosystem) to the molecular level [4]. Nutrient uptake and their internal utilization efficiencies are the two central cores for improving crop NUE [5]. This can be achieved through optimizing agronomic strategies (soil-rhizosphere management) and breeding nutrient-efficient cultivars. Plant genetics and physiological mechanisms and their interaction with best agronomic practice are also a tool that can be used to increase efficiency of cropping systems [3]. Thus, it needs involvement of integrated nutrient management strategies that take into consideration improved fertilizer along with soil and crop management practices are necessary [6]. Sustainable nutrient management must be both efficient and effective to deliver anticipated economic, social, and environmental benefits.

Plants experience nutrient deficiency when soil nutrient availability is either an inherently low amount or low mobility of nutrients in the soil, or poor solubility of certain chemical forms of soil nutrients [7]. Of the various nutrients essential for plants, nitrogen (N), phosphorus (P), and potassium (K) are required in the largest quantities, and their deficiency severely limits crop yield [8]. The dynamic nature of N and P in soil-plant systems creates a unique and challenging environment with nitrate and phosphate contamination of surface and/or groundwater, which can be attributed in large part to low efficiency in plant nutrient uptake. The main challenge for improving P and K use efficiency at the farm level is to apply the existing knowledge in a practical manner [9]. Hence, the best management practice for N, P, and K must consider the specific characteristics of crops, cropping systems, environments, and soils is application of 4R nutrient stewardship. Therefore, this chapter tries to summarize the concept of NUE and recent strategies for enhancing use efficiency of N, P, and K. These approaches consider economic, social, and environmental dimensions essential to sustainable agricultural systems and afford a suitable context for specific NUE indicators.

2. Concept of phosphorus use efficiency (PUE)

The variations in defining nutrient efficient plants and methods used in calculating nutrient use efficiency make it difficult to compare results of different studies [10–13]. Understanding the terminology and the context in which it is used is critical to prevent misinterpretation and misunderstanding and determination of NUE in crop plants is an important approach to evaluate the fate of applied chemical fertilizers and their role in improving crop yields. In order to develop a common framework for NUE, scientists started to formulate concepts and definitions that should serve as a basis for comparison and discussion of research. Nutrient use efficiency in its broadest sense indicates how effectively a plant is able to capture and utilize nutrients to produce biomass. It is simply a measure of how well plants use the available mineral nutrients [10]. The earlier definition of NUE by [14] is simply increment of yield per applied nutrient (Eq. (1)).

$$NUE = \frac{\text{Total productivity } (\text{g cm}^{-2} \text{ year}^{-1})}{\text{Rate of resource uptake or acquisition by plant } (\text{g N m}^{-2} \text{ year}^{-1})} \quad (1)$$

While the most recent and complicated one used in crop modeling formula is (Eq. (2)) [12].

$$NUE = \frac{(R_{ac} - R_{\min}) * NPP_{\max}}{(R_{ac} - R_{\min} + \alpha) * N_{ac}} \quad (2)$$

where $R_{\min}$ is the estimated minimum resource requirement for positive growth, N_{ac} is the amount of nutrient uptake by plant, $NPP_{\max}$ is the production asymptote, and $\propto$ is the half-saturation constant with respect to resource.

Generally, nutrient use efficiency comprises both yield as a function of inputs and percentage of nutrient recovered respectively, contributing to yield and quality [15]. The NUE is based on (a) uptake efficiency (acquire from soil, influx rate into roots, influx kinetics, radial transport in roots are based on root parameters per weight or length, and uptake is also related to the amounts of the particular nutrient applied or present in soil), (b) incorporation efficiency (transports to shoot and leaves are based on shoot parameters), and (c) utilization efficiency (based on remobilization, whole plant, i.e., root and shoot parameters) [4].

Phosphorus use efficiency can be divided into (i) P acquisition efficiency [the capacity of a cultivar to extract P from soil] and (ii) P internal utilization efficiency [the capacity of a cultivar to transform the acquired P into biomass/grain yield] [16–18].

i. Phosphorus uptake or acquisition efficiency (PACE)

   Uptake efficiency or the ability of the plant to extract the nutrient from the soil is calculated as [19] (Eq. (3)).

$$\text{P-uptake efficiency(PACE)} = \frac{\text{Total above} - \text{ground nutrient(P) in the plant at maturity(Nt)}}{\text{Nutrient(P) supplied(Ns)}} \quad (3)$$

ii. Phosphorus utilization efficiency (PUTE)

   Phosphorus utilization efficiency is defined as a crop's ability to convert the absorbed P into grain yield [19] (Eq. (4)) can be calculated as:

$$\text{P-utilization efficiency(PUTE)} = \frac{\text{Total above} - \text{ground plant dry weight at maturity(Tw)}}{\text{Total aboveground plant nutrient at maturity(Nt)}} \quad (4)$$

Utilization efficiency can also be calculates as suggested by [20], (Eqs. (5) and (6)) and expressed as follows:

$$\text{Utilization efficiency} = \text{Harvest index} \times \text{Nutrient biomass production efficiency} \quad (5)$$

$$\text{Utilization efficiency} = \text{Harvest index} \times \text{Inverse of total nutrient concentration in the plant.} \quad (6)$$

Generally, if P supply is limited or in more acidic and calcareous soil, P acquisition could be more important than P utilization and high fertilizer application necessary in order to provide sufficient plant-available P. On the other hand, with adequate P supply, PUTE could be considered more important than PACE for crop P efficiency [17]. Therefore, the improvement of both PACE and PUTE in the given species under different P supply conditions in the different soil types seems to be the perfect breeding approach (**Figure 1**) [17].

Hence, Nutrient use efficiency = Uptake efficiency × Utilization efficiency. All unit dry weights are in g m^{-2} [19].

For nitrogen use efficiency in their various definitions and components (**Figure 2**) [21].

Apparent recovery efficiency is one of the more complex forms of nutrient use efficiency (NUE) expressions and is most commonly defined as the difference in nutrient uptake in above-ground parts of the plant between the fertilized and unfertilized crop relative to the quantity of nutrient applied. It is often the preferred NUE expression by scientists studying the nutrient response of the crop [22]. Reference [23] proposed that the balance method be used to assess fertilizer P efficiency (Eq. (7)). The balance method is described mathematically as:

$$\text{P-use efficincy}(\%) = \frac{\text{P taken up by crop (fertilized soil)}}{\text{Amount of P applied}} \times 100 \quad (7)$$

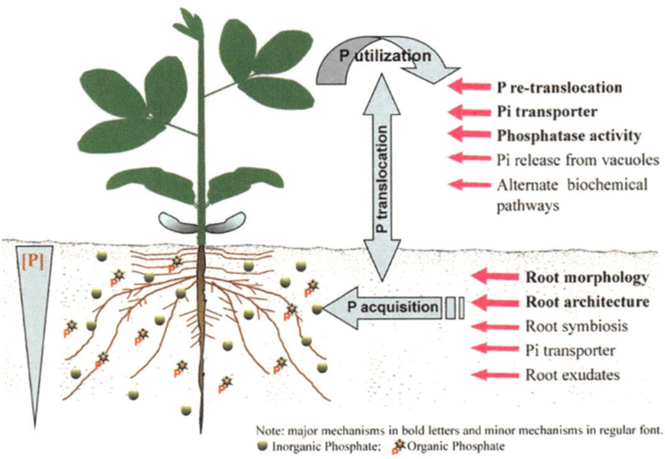

Figure 1.
Schematic representation of the possible mechanisms of P acquisition and utilization for better growth of modern crops grown in intensive cropping systems (adopted from [17]).

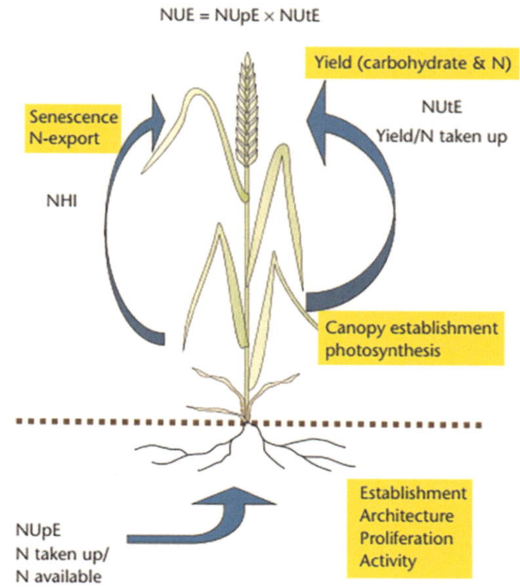

Figure 2.
Illustration of nutrient use efficiency parameters exemplified by NUE in wheat. Key process contributing to the NUE trait: nitrogen uptake efficiency, NUpE; nitrogen utilization efficiency, NUtE; nitrogen harvest index, NHI (adopted form [15]).

3. Why we worry of phosphorus use efficiency

Phosphorus use efficiency has become burning issues in recent times due to several reasons [24]. Unlike N, the amount of P is less-abundant, finite resource, less-available, and poor mobility in the soil, being one of the most inaccessible elements for plants. Its deficiency is a major constraint to agricultural production, and it affects an area of over 2 billion hectares worldwide that is on about 70% of the world's arable land [25]. Remarkably, usually only about 10–30% of the P fertilizer applied in the first year is taken up by the roots, with a substantial part accumulated in the soil as residual P not readily available for plants [26]. This may be due to nature of P that can bound to calcium in alkaline soils and readily complexed to charged Al and Fe oxides and groups hydroxyls on clay surfaces in acidic soils [23]. In addition, agricultural phosphorus (P) run-off is a primary factor in the eutrophication of aquatic and marine ecosystems and has also led to blooms of toxic cyanobacteria [27] and can contain heavy metals such as cadmium that may accumulate in arable soils. Moreover, organic material present in the soil (e.g., from manure or crop residues) can also bind phosphate ions as well as phytate (inositol compounds). In order to avoid a future food-related crisis, phosphorus scarcity needs to be recognized and addressed in contemporary discussions on global environmental change and food security, alongside water, energy, and nitrogen [28].

4. Breeding or crop modification strategies

4.1 Selection of improved genotypes

Selection and breeding nutrient-efficient species or genotypes within a species are justified in terms of reduction in fertilizer input cost of crop production and also reduced risk of contamination of soil and water. Many plants have evolved morphological, physiological, biochemical, and molecular adaptive systems to cope with P-deficiency stress, such as altered root architecture to explore more soil volume and increased carboxylate exudation containing phosphatases, nucleases, and various organic acids [29]. These mechanisms and strategies are necessary to liberate or solubilize Pi from organic and other insoluble pools [30], enhance Pi uptake capacity [31], recycle internal Pi remobilize/retranslocate P from mature to young developing organs [32, 33], and reprioritize metabolic P utilization [34]. Under the current situation, farmers need P-efficient genotypes that perform better than other genotypes with equivalent P inputs. Therefore, selection/identification of cultivars that can absorb and use P efficiently is a promising strategy to cope with environments deficient in bio-available P. Due to the diverse functional and structural roles of P in plants, P-use efficiency (PUE) is a complex trait to dissect [24].

4.2 Modification of root morphology and physiology

The root morphological factors such as length, thickness, surface area, and volume have profound effects on the plant's ability to acquire and absorb nutrients in soil [35]. These parameters are influencing the ability of the roots to penetrate high density soil layers, to extremes tolerate temperature, moisture, toxicities, and deficiencies of elements. Additionally, they have the ability to modify the rhizosphere pH and the nutrient uptake kinetics. Efficient acquisition will depend first on root architecture in terms of transporters and exudates and often the presence of symbiotic associations such as mycorrhiza. Hence, improving early root establishment, high-affinity transporter systems, association of microorganisms (mycorrizha), proliferation of roots, and enhanced mechanisms for increasing bio-availability of nutrients and then enhancing NUE [5]. Improvement of transporters plays essential roles particularly in conjunction with effective root proliferation in contributing to nutrient use efficiency. The other important attribute for uptake efficiency is having adequate sinks to store acquired nutrients, which will prevent negative feedback regulation on the initial acquisition/assimilatory processes and should provide important remobilizable storage [5]. The second component of uptake efficiency is root physiological activity such as differing uptake kinetics, i.e., maximum net influx (I_{max}), affinity of the transporter (K_m) and the roots depletion ability (C_{min}), which result in different nutrient uptake rates per unit root and time due to their effect on P diffusion [36]. Lower Km values (higher affinity) and higher I_{max} values indicate a higher uptake rate of plants for a determined nutrient at low concentration [11].

A recent study further showed that root tips also play an important role and, despite their small size, accounted for approximately 20% of the total seedling Pi uptake [37], mainly increasing organic acid exudation strategies [38]. Plants increase total soil exploration by increasing root length, increasing root branching, increasing specific root length (i.e., roots with smaller diameter), and modifying branching angle [39–41]. The findings of Bates and Lynch [39] suggested that increased root growth is associated with improved plant performance under low P by exploring a larger volume of soil. Consequently, root: shoot ratio increases significantly in low-P environments and is an excellent index for partitioning photosynthesized carbon between above- and below-ground plant parts. Root density and root: shoot ratio generally increased under P deficiency, thus favoring P acquisition by plants [29].

Genetic variation for root hair traits, particularly root hair length, can be exploited in breeding for improved P uptake efficiency and P fertilizer use efficiency in crops.

Moreover, a deeper root with more aerenchyma tissues in the cortex of the roots can also be an important trait that contributes to efficient N uptake with lower carbon input in root growth [42]. This root architecture may also be efficient in the uptake of deep water and therefore help to increase drought resistance [43]. However, Miguel et al. [44] showed in field trials that shallow and hairy root traits are synergistic in their effects on Pi uptake by bean. However, modifying root growth in response to nutrient deficiency, it is a challenge and complex to identify key regulators that are sufficiently upstream and robust to be suitable for developing plants with optimized root systems for nutrient uptake [8].

4.3 Improving translocation (partitioning/remobilization)

Levels of fertilizer applications influence the total dry matter accumulation, thereby affecting the nutrient demand (uptake/utilization) [9]. Improved nutrient utilization efficiency from agrochemicals through PGPR and (or) AMF can contribute to the protection of water resources against agro-pollution and reduce the growing cost of fertilizers [10]. After inorganic phosphate (Pi) acquisition from rhizosphere, Pi should be efficiently transported to shoot for the requirement of plant growth by phosphate transporters (*Pht1, Pht2, Pht3,* and *Pht4*), which are located on the plasma membrane, plastidial membrane, mitochondrial membrane, and Golgi compartment, respectively [45]. In crops, a large fraction of the Pi present in vegetative organs is remobilized to the grain during the reproductive growth, and soil Pi availability at this stage has a relatively small effect on grain yield. Enhanced expression of high-affinity, plasma-membrane-bound Pi transporters in roots and a concomitantly increased P-up- take capacity were reported as a typical P-starvation response [46]. Moreover, enhanced metabolic activities of young tissues make them stronger sinks for the already absorbed P. Remobilization of stored P in the stem and older leaves to metabolically active sites may supplement the restricted P supply under P deficiency [29].

Another promising area for improvement of crop NUE is to enhance the efficiency of nutrient remobilization from senescing organs to young, developing organs, particularly immature leaves, and developing seeds [47]. The senescence process, that is, the dying-off of vegetative plant parts during seed maturation, is at the core of the nutrient use efficiency issue, as the nutrients need to be remobilized from these parts and translocated into the developing seed [48]. Maximizing the effectiveness of P-remobilization from senescing organs could make an important contribution to the development of crops that can tolerate Pi deficiency, because senescing organs of most "modern" crop varieties exhibit low P-remobilization efficiencies of <50% [30]. An integral understanding of P remobilization would facilitate development of effective biotechnological strategies to improve crop PUE, thereby reducing the rate of depletion of nonrenewable rock P reserves [30, 47]. Therefore, mobilization and redistribution of P from the old tissues to the young tissues will also contribute to high P use efficiency. Better distribution of nutrients in parts of plant (root, shoot, and grain) reflects their use efficiency [11].

4.4 Improving internal utilization

In the plant, uptake and utilization efficiency of nutrients are governed by different physiological mechanisms and their response to deficiency, tolerance, and toxicity of element(s) and climatic variables [49]. Efficient internal utilization of nutrient is generally attributed because of high photosynthetic activity per unit of nutrient (P) and more efficient P remobilization from older to young leaves [47]. Acid phosphatase contributes to the increased P utilization efficiency in bean through P remobilization from old leaves [50]. Therefore, improving higher total chlorophyll concentration [51], enhancing phosphorylase stimulation [52], and improving partitioning of carbon between glycolytic and pentose phosphate pathways [53] also provide an effective approach to improve phosphorus use efficiency and crop productivity simultaneously.

P-utilization efficient cultivars produce high yield per unit of absorbed P under P deficient conditions, since they have low internal P demand for normal metabolic activities and growth. Hence, they have low requirement for mineral P fertilizer inputs to produce reasonably high yield. Moreover, they remove less P from soil during growth and therefore the quantity of P removed along with the harvestable parts of the crop would obviously be less, consequently reducing the quantity of mineral P fertilizer inputs required for maintenance fertilization [54].

5. Optimization of agronomic practice

Agronomic practices can change soil physicochemical properties and biological characteristics. As a result, a number of agronomic practices have been proposed to enhance nutrient availability under diverse climatic conditions [55, 56]. The rhizosphere (root-soil interface) is the most important area for plant–soil-microorganism interactions and is the hub for controlling nutrient transformation and plant uptake [7]. This modification is paramount to increase nutrient availability and to minimize losses in surface runoff. Possible management strategies options for improving NUE through optimizing agronomic practice or rhizosphere modification [57] are the following:

5.1 Application of nutrient stewardship concept

The 4R Nutrient Stewardship framework promotes the application of nutrients using the right source (or product) at the right rate, right time, and right place. The framework was established to help convey how fertilizer application can be managed to ensure alignment with economic, social, and environmental goals [58]. Nutrient Stewardship defines the right source, rate, time, and place for fertilizer application as those producing the economic, social, and environmental outcomes desired by all stakeholders of the plant ecosystem (**Figure 3**). This 4R techniques applies (1) right rate—supplying growing crops with the right amount of nutrients for healthy growth and development based on experimentation under various environmental conditions; (2) right time—matching nutrient availability to with the timing of plant peak nutrient uptake and demand; (3) right placement adding nutrients to the soil at a place where crops can easily access them related to volume of roots.; (4) right source—applying the correct fertilizer and organic resources that provide growing crops with all nutrients required for good growth and maturity [58]. The 4R concept was established to help convey how fertilizer application can be managed to ensure alignment with economic, environmental, and social goals [22, 59].

Soil testing remains one of the most powerful tools available for determining the nutrient supplying capacity of the soil, but to be useful for making appropriate fertilizer recommendations good calibration data is also necessary [2]. As P is less mobile, less soluble, and highly prone to soil fixation; effectiveness of applied P depends on the properties of soil being fertilized, fertilizer itself, and time and method of its application [60]. To enhance phosphorus use efficiency (PUE) of applied P fertilizer,

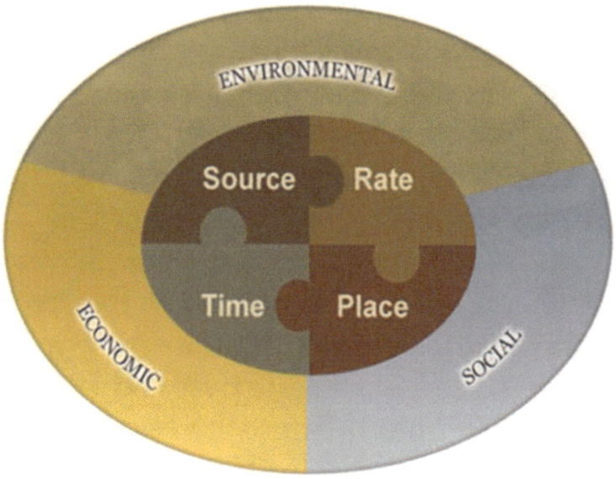

Figure 3.
The 4R nutrient stewardship concept (adopted from [22, 59]).

time and method of its application are critically important, because different P application methods differ in PUE [61]. In highly sandy soils, P may need to be managed like N, by splitting applications and applying small amounts at sowing and topdressing later in the crop growth cycle [62]. Studies of Jing et al. [63] suggested that localized supply of superphosphate combined with ammonium-N (NH_4^+-N) significantly stimulated root proliferation, especially of fine roots, and thus improved maize growth in a calcareous soil. Further studies indicated that localized supply of P and NH_4^+-N at both seeding and later growth stages increased maize yield by 8–10%, P uptake by 39–48%, and localized increases in root density and length of 50% [64]. Rehim et al. [65] also reported that the fixation of broadcasted P is much greater than the fertilizer applied in bands because of less contact with P fixing ingredients. At higher P application, the adsorption of P increased because the plants readily utilize only 8–33% of applied P in the first growing season and remaining portion remained fixed that consequently resulted in higher Olsen P. So, at higher P application rates, plants used smaller proportion of fertilizer P that resulted in low PUE [61].

In principle, N deficiency increases root growth, resulting in longer axial roots (primary roots, seminal roots, and nodal roots), and this helps maize roots to explore a larger soil volume and thus increases the spatial N availability [66]; however, long-term N deficiency stunts root growth due to insufficient N. But also, root elongation can be inhibited if the N supply is too high. Excessive application of N-P fertilizers may lead to high concentrations of soluble nutrients in the root zone, which can also restrict root growth and rhizosphere efficiency [67], even small amounts of P lost can be a cause of the adverse effects of eutrophication of surface waters. Therefore, judicious application of fertilizer best management practices (BMP) [22] that includes the right rate [68], right time [69], right source, right place, and balanced fertilization (4RB) is the best management practice for achieving optimum nutrient efficiency [2, 22].

5.2 Cereal-legume intercropping

Cereal-legume intercropping is a crop production system utilized to improve productivity and sustainability under diverse environmental conditions. It can also improve nutrient use efficiency and crop productivity [7]. Intermingling of maize and faba-bean roots increased N acquisition by both crop species by about 20% compared with complete or partial separation of the root systems. Further studies indicate that N_2 fixation can be improved by yield maximization in the intercropping system. The improved productivity observed in this production system has been associated with increased levels of available phosphorus (P) in the root rhizosphere. Hinsinger et al. [70] reported more stable yield, superior land resource utilization or conservation, and enhanced pest or weed control [71–73]. Furthermore, cereal-legume intercropping can also enhance the phosphatase enzyme activity and available P in the soil due to rhizosphere acidification by the legumes in the cropping system [74].

The possible mechanism that increases PUE in intercropping is the increased rhizosphere soil acid phosphatase (RS-AP_{ase}) activity observed in intercropping due to the fact that large amounts of acid phosphatase are known to be released from their roots into the root rhizosphere. The (RS-AP_{ase}) activity was significantly higher (26–46%) in the intercropping and occurred concomitant with a significant increase in available phosphorus (RS-$P_{available}$) in the rhizosphere on podzols in cool climate boreal ecosystem [75]. Another mechanism could be secreting H^+ into the soil that acidifiies the rhizosphere [57, 76] and improves dissolution of phosphorus and then enhances P-availability [70]. Additional possible mechanism that improves of plant growth and P uptake in mixed planting was due to root interspecific complementation or facilitation. The complementarity between root morphological and physiological traits of neighboring plants underpins the interactive facilitation, which was the main underlying mechanism improving nutrient-use efficiency, particularly of P, in mixed cropping system [77, 78]. The complementary niches of maize and faba bean significantly reduce interspecific nutrient competition and thus improve nutrient-use efficiency [79]. The presence of maize increased the secretion of carboxylates from alfalfa roots, suggesting that the root interactions between maize and alfalfa are crucial for improving P-use efficiency and productivity in intercropping [80]. Subsequently, Sun et al. [76] reported that decreasing rhizosphere pH and increasing organic anion exudation played key roles in soil P mobilization of maize and alfalfa, with little contribution of acid phosphatase.

5.3 Effective microbial inoculation

The mycorrhizal symbiosis particularly, arbuscular mycorrhizal fungi (AMF), is arguably the most important symbiosis on earth [81]. AMF colonize the roots of many agriculturally important food and bioenergy crops form (approximately 80–90% of all known land plant species) [81] and could serve as "biofertilizers and bioprotectors" in environmentally sustainable agriculture [82]. In AMF associations, two pathways for plant P uptake exist: the direct pathway (P uptake by roots) and the AM fungal pathway [83]. AMF facilitates the uptake and transfer of mineral nutrients, such as phosphorus, nitrogen, sulfur, potassium, calcium, copper, and zinc, from the soil to their host plants by means of the extraradical mycelium extending from colonized roots into the soil [84]. The contribution of AMF to P uptake reaches up to 77% under low P supply compared with only 49% under high P supply [85]. Furthermore, the commercial inoculum Mycobiol, consisting of Glomus spp., *Entrophospora colombiana, and Acaulospora mellea*, enhanced P acquisition and plant growth in a pot experiment [86]. González and Walter [87] observed that *Glomus aggregatum* increased P uptake and biomass production of inoculated plants compared.

Various mechanisms have been suggested for the increase in the plant uptake of P. These include: exploration of larger soil volume; faster movement of P into mycorrhizal hyphae; and solubilization of soil phosphorus [88]. Exploration of larger soil volume by mycorrhizal plants is achieved by decreasing the distance that P ions must diffuse to plant roots and by increasing the surface area for absorption. Faster movement of P into mycorrhizal hyphae is achieved by increasing the affinity for P ions and by decreasing the threshold concentration required for absorption of P [88]. Solubilization of soil P is achieved by rhizospheric modifications through the release of organic acids, phosphatase enzymes, and some specialized metabolites such as siderophores [55].

The composition and amount of root exudates affect the composition of microbes in the rhizosphere and the structure of the rhizosphere microbiome, affecting plant growth and nutrient uptake [81]. For precision rhizosphere management, plant-microbe interactions must be finely tuned to improve P use efficiency by crops [57]. **Figure 4** illustrates the main structural differences between AM (more for P absorption) and ectomycorrhizal (more for N and few for P absorption) associations of angiosperms or gymnosperms [81].

Among the soil bacterial communities, ectorhizospheric strains from Pseudomonas and Bacilli and endosymbiotic rhizobia have been described as effective phosphate solubilizers [90]. Phosphate-solubilizing bacteria (PSB) are also capable of making P available to plants from both inorganic sources and organic ones and increasing P-fertilizer-use efficiency by different mechanisms [91]. They are rhizobacteria that convert insoluble phosphates into soluble forms through acidification, chelation, exchange reactions, and the production of organic acids [92]. Therefore, combined application of AMF and P solubilizers [93] and N fixers are the best inoculants. AM fungi together with PSMs could be much more effective in supplementing soil P. Understanding AM-plant symbiosis, developing AM fungi that could be cultured in vitro, and developing P-solubilizing AM will help realize their potential as phosphate biofertilizer [94].

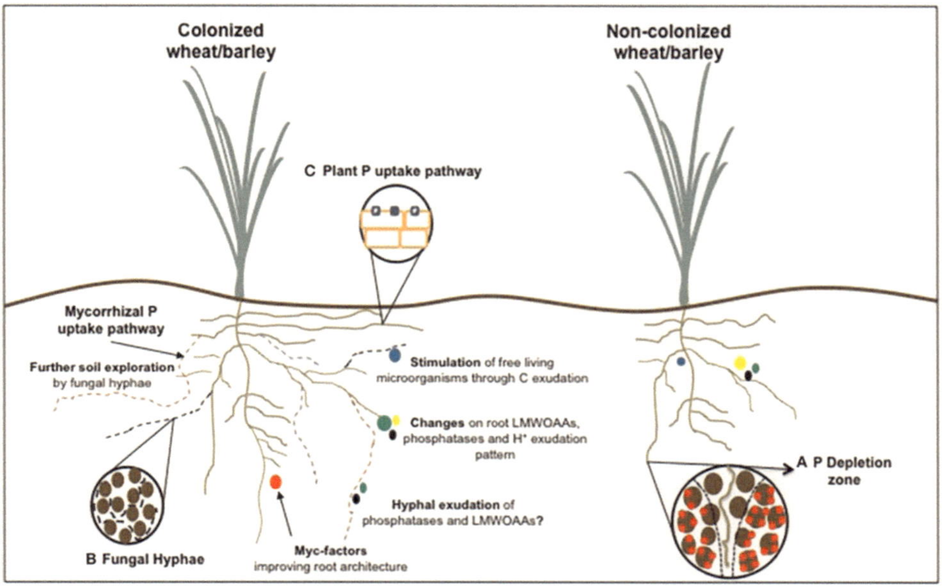

Figure 4.
Phosphorus acquisition efficiency related traits of wheat and barley roots affected by arbuscular mycorrhizal symbiosis in comparison to a non-colonized counterpart (adopted from [89]). (A) Representation of P depletion zone around the rhizosphere; (B) access to smaller soil pores by AM fungal hyphae; and (C) modulation of plant P transporters following colonization.

5.4 Regulating soil pH

Soil pH is one of the most important chemical properties influencing nutrient solubility and hence availability to plants. Large amount of P applied as fertilizer enters in to the immobile pools through precipitation reaction (fixation) with highly reactive Al^{3+} and Fe^{3+} in acidic and Ca^{2+} in calcareous or normal soils [94]. Acidic, highly weathered, iron (Fe)-rich soils rapidly bind phosphates at mineral surfaces, limiting access to plant roots. Furthermore, applied Pi (inorganic P) is quickly fixed into insoluble inorganic or organic forms due to its high reactivity and microbial action [95].

Soil pH markedly limits plant growth and P chemistry in soils through its effect on P adsorption and through interactions that affect precipitation of P into solid forms in soil [62]. Consequently, about 80–90% soil P becomes unavailable depending on soil composition and pH [96], 50–70% of the total applied conventional fertilizers are lost to the environment. This level of loss in agricultural nutrients not only leads to the loss of valuable resources but also causes the severe reduction of yield [97]. The pH of a calcareous soil is reduced by the presence of gypsum ($CaSO_4{\cdot}2H_2O$) due to the concentration of Ca^{2+}, which would be expected to decrease the sorption of P, if followed by leaching to removed much of the soluble Na^+ and Ca^{2+} [98]. Thus, adjusting soil pH and base saturation are methods to reduce the amount of P that is bound by Al, Fe, and Ca, further reducing the effects of Al toxicity to plants, which can inhibit uptake, and use of P by the plant (**Figure 5**) [23, 99].

Lime acidic soil is widely used in agriculture to create and maintain a soil pH optimal for plant growth in acid soils. Lime reduces toxic effects of hydrogen, aluminum, and manganese, improves soil biological activities, cation exchange capacity (CEC), P, Ca, and Mg availability and soil structure, promotes N_2 fixation, stimulates nitrification, and decreases availability of K, Mn, Zn, Fe, boron (B), and Cu [11]. An increase in soil pH, as a result of liming, was due to an increase in hydroxide ions, which increases microbial activity and communities, hence, increasing decomposition of soil organic matter and release of Fe and Al [100]. The decrease in Al-P and Fe-P could be due to their precipitation as insoluble $Al(OH)_3$ and $Fe(OH)_3$ after increased addition of liming material [101]. In addition, Al and Fe oxides become more negatively charged with an increase in pH contributing to an increase in available P [102].

Liming, gypsum application, or mixing of both is an effective practice to improve pH, improve Ca content, and control Al toxicity. Lime has very low mobility in soil, and when surface applied, it does not reduce the acidity of subsurface soil horizons. Contrary to lime, gypsum ($CaSO_4$) has a greater downward movement, and when applied to the surface, it can still impact and reduce the acidity of the subsoil [4]. The pH of a

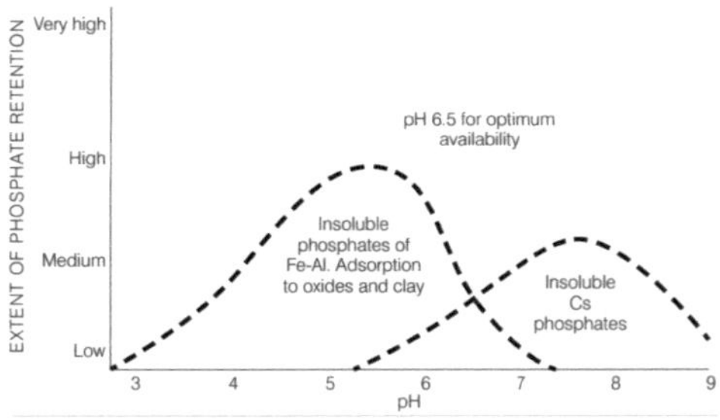

Figure 5.
Soil P availability as affected by soil pH (adopted from Havlin et al. 1999).

calcareous soil is reduced by the presence of gypsum ($CaSO_4{\cdot}2H_2O$) due to the concentration of Ca^{2+}, which would be expected to decrease the sorption of P, if followed by leaching to removed much of the soluble Na^+ and Ca^{2+}. The uptake of nutrients by plants, content of nutrients in plants and in soil were substantially positively influenced by both the wood ash, especially by FGD gypsum [103]. Gypsum application can ameliorate saline-sodic soil, thereby increasing crop yield and NUE [104].

5.5 Application of advanced techniques

Apart from traditional methods, new techniques have been developed such as site-specific/real-time nitrogen management, slow release/controlled release fertilizer (SR/CRF), site-specific precision nutrient management, and urease/nitrification inhibitor. Those techniques play an important role in decreasing fertilizer loss and increasing NUE [105]. The remote sensing is quicker than the previous two methods, and it obtains continuous data rather than spot data, which is more advantageous. It is becoming the major means of obtaining data for precision farming. GIS (geographic information system) establishes the field management information system by processing, analyzing, and trimming the data of soil and crops [105]. Another approach to synchronize release of N from fertilizers with crop need is the use of N stabilizers and controlled release fertilizers. Nitrogen stabilizers (e.g., nitrapyrin, DCD [dicyandiamide], NBPT [n-butyl-thiophosphoric triamide]) inhibit nitrification or urease activity, thereby slowing the conversion of the fertilizer to nitrate. The most promising for widespread agricultural use are polymer-coated products, which can be designed to release nutrients in a controlled manner.

Agronomic management strategies such as precision P fertilization, polymer coated P-fertilizers, and recycling of P from domestic, agricultural, and industrial wastes can be helpful in improving P use at farm level [106]. Modern concepts for tactical N management should involve a combination of anticipatory (before planting) and responsive (during the growing season) decisions [9]. On soils with moderate P and K levels and little fixation, management must focus on balancing inputs and outputs at field and farm scales to maximize profit, avoid excessive accumulation, and minimize risk of P losses. Improving the internal, on-farm and field recycling is the most important K management issue worldwide. As for N, the primary determinants for RE_P and RE_K are the size of the crop sink, soil supply, soil characteristics, and fertilizer rate.

Control release fertilizers with polymer coatings are commonly applied to crops to increase efficiency of nutrients [96]. One way of improving the P availability to crop plant is by coating diammonium phosphate (DAP) with polymer that allows a steady but controlled discharge of phosphorus from the granules for crop plant uptake and improved P recovery percentage. Thus, by the use of polymer, availability of P to plant increased because it has high cation exchange capacity, which holds the divalent calcium (Ca^{+2}) and trivalent cations iron and aluminum (Fe^{+3} and Al^{+3}) and stop P fixation with these cations. Moreover, polymer absorbs water efficiently and holds more water and keeps P in available form that enhanced the plant growth and yield-contributing factors [97]. This is because polymer-coated diammonium phosphate (DAP) absorbs water many times of its original weight, which increases the availability of phosphorus for longer period of time [107] and creates a diffusion shell around the grain of DAP and directly reduces the fixation and precipitation by reducing the availability of calcium and magnesium (Ca^{+2}/Mg^{+2}) cations [108]. As the result of this mechanism, availability of phosphorus to plants increases and leads to more P uptake, and this uptake indirectly influences the other nutrient absorption by crop plants.

5.6 Use integrated soil nutrient management practice

Considering the wide variety of soil types, cropping patterns, and farmers' resources, several management practices are adopted to reduce the magnitude of soil fertility degradation. Integrated Plant Nutrient Management System (IPNMS) is defined as the package of practices for the manipulation of the plant growth environment to supply essential nutrients to a crop in an adequate amount and proportion for optimum production without degrading the natural resources [3]. Many authors have reported that combining organic and inorganic P can improve and sustain crop yields in low fertility soils [109–111]. Best management practices (BMPs) such as use of fertilizer and amendment (lime), proper crop rotations, increases in organic matter content, and control of erosion, insects, diseases, and weeds can significantly improve crop yields and optimize nutrient use efficiency [11]. Integrated use of organic manures and fertilizers not only improves efficiency of crops but also significantly increases the availability of P [112, 113].

Organic amendment improves the structure and fertility of the soil by adding nutrients and organic matter and consequently promotes soil microbial biomass and activity. Blockage of P sorption sites by organic acids, as well as complexation of exchangeable Al and Fe in the soil, is potential cause of this mobilization [114]. Organic materials can reduce P fixation by masking the fixation sites on the soil colloids and by forming organic complexes or chelates with Al, Fe, and Mn ions, thereby improving P uptake efficiency of crop plants. Decomposition of organic matter produces organic anions that interact with soil to reduce P sorption via (1) complexation/competition for soil P binding sites such as Fe and Al oxyhydroxides or (2) increased soil P^H. Organic materials also increase agronomic efficiency by improving availability of P by promoting soil aggregation, increased soil P^H, microbial biomass, and parameters controlling soil-P-sorption [115]. The integration of biochar FYM, poultry manure, and inorganic P sources increases in PUE under both wheat and maize crops, and there is a concomitant increase in crop yields compared with the unamended soil [112, 113]. This increase in PUE with biochar addition could also be the result of the additional nutrients made available by biochar [112]. Similarly, FYM applications increase soil P bioavailability more than applications of triple supper phosphate that enhance P Uptake Efficiency. FYM is also a source of other nutrients used by crops via mineralization, which promotes root development and root area interception and thus increases nutrient uptake including P uptake [116].

Rotating a legume with a cereal can enhance P acquisition by cereals through indirect feedback interactions [117]. A legume crop modifies the rhizosphere through biological and chemical processes, thereby increasing P uptake by the following cereal crop. As reported by [77], legumes are able to mobilize P that is not initially available to cereal species, thereby improving the availability of P for the following crop. The biological processes include the promotion of symbiotic mutualists such as nitrogen-fixing rhizobacteria and mycorrhizal fungi, while the chemical processes are acidification of the rhizosphere and secretion of organic anions [79].

6. Conclusion

Achievement and maintenance of high nutrient use efficiency (NUE) together with high crop productivity have become a major challenge in both developed and developing countries with an increasing growing population, depletion of natural resources, and deteriorating environmental conditions. Improving nutrient use efficiency (NUE) in plants is vital to enhance the yield and quality of crops, reduce nutrient input cost and improve soil, water, and air quality [3]. Higher NUE by plants could reduce fertilizer input costs, decrease the rate of nutrient losses, and enhance crop yields. Improving crop nutrient use efficiency ideally requires an understanding of the whole system, from the macro (agro-ecosystem) to the molecular level.

The development of nutrient-efficient crop varieties that can grow and yield better with low supply is a key to improving crop production. A prerequisite for nutrient use efficiency for any germplasm will be the optimization of agronomic practice for any given environment and season. Judicious application of fertilizer that includes the right rate, right time, right source, right place, and balanced fertilization (4RB) is the best management practice for achieving optimum nutrient efficiency. By the coordination of the acquisition, root-to-shoot translocation, utilization, and remobilization of internal Pi can be achieved through genetic breeding. Selection and breeding nutrient efficient species or genotypes within a species are justified in terms of reduction in fertilizer input cost of crop production and also reduced risk of contamination of soil and water. Overall NUE in plant is a function of capacity of soil to supply adequate levels of nutrients and ability of plant to acquire, transport in roots and shoot, and remobilize to other parts of the plant. Improvement in NUE will ultimately come from integrating a range of different approaches to develop a more efficient farming system. Use of nutrient efficient crop species or genotypes within species in combination with other improved crop production practices offers the best option for meeting the future food requirements of expanding world populations. Modern tools and resources available to plant scientists and the agronomy and breeding communities should aid further improvements in NUE and hence crop production. Therefore, integrated strategy that seeks to increase phosphorus use efficiency and simultaneously seeks to recover unavoidable phosphorus losses. The nutrient inputs in the intensive farming system should be optimized to achieve both high crop productivity and high nutrient use efficiency through maximizing root/rhizosphere efficiency in nutrient mobilization and acquisition.

Acknowledgements

The authors are highly thankful to researchers whose findings are included directly or indirectly in preparing this manuscript.

Funding

The authors received no direct funding for this research.

Conflicts of interest

The authors declare no conflict of interest.

Data availability

All data generated are included in this article reference's part.

Abbreviations

AMF	arbuscular mycorrhizal fungi
BMP	best management practice
DAP	diammonium phosphate
FYM	farm-yard manure
NUE	nutrient use efficiency
PACE	phosphorus acquisition efficiency
PSB	phosphate solubilizing bacteria
PUE	phosphorus use efficiency
PUTE	phosphorus utilization efficiency

Author details

Addisu Ebbisa
Faculty of Agriculture, Department of Plant Science (Agronomy), Haramaya University, Harar, Ethiopia

*Address all correspondence to: saebbisa@gmial.com; aebbisa@gmail.com

References

[1] Salim N, Raza A. Nutrient use efficiency (NUE) for sustainable wheat production: A review. Journal of Plant Nutrition. 2019;**43**(2):1-19. DOI: 10.1080/01904167.2019.1676907

[2] Roberts TL. Improving nutrient use efficiency. Turkish Journal of Agricultural. 2008;**32**:177-182

[3] Baligar VC, Fageria NK, He ZL. Nutrient use efficiency in plants. Communications in Soil Science and Plant Analysis. 2001;**32**(7/8):921-950

[4] Baligar VC, Fageria NK, He ZL. Communications in soil science and plant analysis nutrient use efficiency in plants. Communications in Soil Science and Plant Analysis. 2016;**32**(7–8):921-950. DOI: 10.1081/CSS-100104098

[5] Hawkesford MJ. An overview of nutrient use efficiency and strategies for crop improvement. In: First PB, Hawkesford MJ, editors. Generic Aspects of Crop Nutrition. Hoboken, New Jersey, United States: John Wiley & Sons, Inc; 2011. pp. 1-16

[6] Fageria NK, Baligar VC. Enhancing nitrogen use efficiency enhancing nitrogen use efficiency. Advances in Agronomy. 2005;**88**(05):98-185. DOI: 10.1016/S0065-2113(05)88004-6

[7] Zhang F, Shen J, Zhang J, Zuo Y, Li L, Chen X. Rhizosphere Processes and Management for Improving Nutrient Use Efficiency and Crop Productivity: Implications for China. 1st ed. Vol. 107 (10). Amsterdam, Netherlands: Elsevier Inc; 2010

[8] Wissuwa M, Heuer S, Gaxiola R, Schilling R, Herrera-estrella L, Damar L. Improving phosphorus use efficiency: a complex trait with emerging opportunities. The Plant Journal. 2017, 2018;**90**(5):868-885. DOI: 10.1111/tpj.13423

[9] Dobermann A. Nutrient use efficiency – measurement and management. Agronomy & Horticulture. 2007;**1442**:1-28

[10] Adesemoye AO, Torbert HA, Kloepper JW. Enhanced plant nutrient use efficiency with PGPR and AMF in an integrated nutrient management system. Canadian Journal of Microbiology. 2008; **886**:876-886. DOI: 10.1139/W08-081

[11] Fageria NK, Baligar VC, Li YC. Journal of plant nutrition. Journal of Plant Nutrition. 2008;**31**(6):1121-1157. DOI: 10.1080/01904160802116068

[12] Alshaal T, El-ramady HR, Al-saeedi AH, Shalaby TA. Essential Plant Nutrients. Berlin/Heidelberg, Germany: Springer; 2017

[13] Dibb BDW. The mysteries of nutrient use efficiency. Better Crops. 2000;**84**(3): 3-5

[14] Berendse F, Aerts R. Nitrogen use efficiency: A biologically meaningful definition. Functional Ecology. 1987;**1** (3):1-32

[15] Hawkesford MJ. Improving Nutrient Use Efficiency in Crops. Chichester: eLS. John Wiley Sons, Ltd; 2012. pp. 1-8. DOI: 10.1002/9780470015902.a0023734

[16] Vance CP, Vance CP, Uhde-stone C, Allan DL. Phosphorus acquisition and use: Critical adaptations by plants for securing a nonrenewable resource. The New Phytologist. 2003;**157**:423-447

[17] Wang X, Shen J, Liao H. Plant science acquisition or utilization, which

is more critical for enhancing phosphorus efficiency in modern crops ? Plant Science. 2010;**179**(4):302-306. DOI: 10.1016/j.plantsci.2010.06.007

[18] Manschadi AM, Kaul H, Vollmann J, Eitzinger J, Wenzel W. Developing phosphorus-efficient crop varieties—An interdisciplinary research framework. Field Crops Research. 2014;**102**:87-98. DOI: 10.1016/j.fcr.2013.12.016

[19] Moll RH, Kamprath EJ, Jackson WA. Analysis and interpretation of factors which contribute to efficiency of nitrogen utilization. Agronomy Journal. 1982;**74**(3):562-564

[20] Ortiz-Monasterio RJI, Rajaram SKDS, McMahon M. Genetic progress in wheat yield and nitrogen use efficiency under four nitrogen rates. Crop Science. 1997;**37**:898-904. DOI: 10.2135/cropsci1997.0011183X003700030033x

[21] Hawkesford MJ, Kopriva S, De Kok LJ. Nutrient Use Effiency in Plants. New York Dordrecht London: Springer; 2014

[22] Roberts TL, Johnston AE. Resources, conservation and recycling phosphorus use efficiency and management in agriculture. Resources, Conservation and Recycling. 2015;**105**:275-281. DOI: 10.1016/j.resconrec.2015.09.013

[23] Syers K, Johnston AE, Curtin D. Efficiency of Soil and Fertilizer Phosphorus: Reconciling Changing Concepts of Soil Phosphorus Behaviour with Agronomic Information. Vol. 18 (108). Rome: FAO; 2008

[24] Bovill WD, Huang CY, Mcdonald GK. Genetic approaches to enhancing phosphorus-use efficiency (PUE) in crops: Challenges and directions. Crop & Pasture Science. 2013;**64**: 179-198

[25] Kirkby EA, Johnston AEJ. Soil and fertilizer phosphorus In relation to crop nutrition. Ecophysiology of Plant-Phosphorus Interactions. 2008;**9**: 177-223

[26] Rose TJ, Wissuwa M. Rethinking Internal Phosphorus Utilization Efficiency: A New Approach is Needed to Improve Pue in Grain Crops Provided for Non-commercial Research and Educational Use Only. Not for Reproduction, Distribution or Commercial Use. 1st ed. Vol. 116. Amsterdam, Netherlands: Elsevier Inc; 2012

[27] Conley DJ et al. Controlling eutrophication: Nitrogen and phosphorus. Ecology. 2009;**323**:1014-1015

[28] Cordell D, Drangert J, White S. The story of phosphorus: Global food security and food for thought. Global Environmental Change. 2009;**19**: 292-305. DOI: 10.1016/j.gloenvcha.2008.10.009

[29] Aziz T, Ahmed I, Farooq M, Aamer MM, Sabir M. Variation In phosphorus efficiency among brassica cultivars I: Internal utilization and phosphorus remobilization. Journal of Plant Nutrition. 2011;**34**:2006-2017. DOI: 10.1080/01904167.2011.610487

[30] Veneklaas EJ et al. Opportunities for improving phosphorus-use efficiency. New Phytologist. 2012;**195**(2):306-320

[31] Neumann GN et al. Physiological aspects of cluster root function and development in phosphorus-de ® cient White lupin (*Lupinus albus* L.). Annals of Botany. 2000;**85**:909-919

[32] Abbas M, Irfan M, Shah JA, Memon MY. Intra-specific variations among wheat genotypes for phosphorus use

efficiency. Asian Journal of Agriculture and Biology (AJAB). 2018;**6**(1):35-45

[33] Irfan M, Abbas M, Shah JA, Akram MA, Depar N, Memon MY. Field study aiming at higher grain yield and nutrient use efficiency in wheat grown in alkaline calcareous soil. An International Journal of Transitional Justice. 2019;7(1):1-9

[34] Aziz T, Lambers H, Nicol D, Ryan MH. Mechanisms for tolerance of very high tissue phosphorus concentrations in Ptilotus polystachyus. Plant, Cell & Environment. 2015;**38**:790-799

[35] Bengough AG, Mckenzie BM, Hallett PD, Valentine TA. Root elongation, water stress, and mechanical impedance: A review of limiting stresses and beneficial root tip traits. Journal of Experimental Botany. 2011;**62**(1):58-68. DOI: 10.1093/jxb/erq350

[36] Steingrobe B, Claassen N. Potassium dynamics in the rhizosphere and K efficiency of crops. Journal of Plant Nutrition and Soil Science. 2000;**163**: 101-106

[37] Kanno S et al. A novel role for the root cap in phosphate uptake and homeostasis. eLife. 2016;**5**(e14577):1-16. DOI: 10.7554/eLife.14577

[38] Delhaize E, Ryan PR, Hocking PJ, Richardson AE. Effects of altered citrate synthase and isocitrate dehydrogenase expression on internal citrate concentrations and citrate efflux from tobacco (*Nicotiana tabacum* L.) roots. Plant and Soil. 2003;**248**:137-144

[39] Bates TR, Lynch JP. Root hairs confer a competitive advantage under low phosphorus availability. Plant and Soil. 2001;**236**(2):243-250

[40] Lynch JP. Roots of the second green revolution. Australian Journal of Botany. 2007;**55**(14):493-512

[41] Gahoonia TS, Nielsen NE. Barley genotypes with long root hairs sustain high grain yields in low-P field. Plant and Soil. 2004;**262**:55-62

[42] Postma JA, Lynch JP. Theoretical evidence for the functional benefit of root cortical aerenchyma in soils with low phosphorus availability. Annals of Botany. 2011;**107**:829-841. DOI: 10.1093/aob/mcq199

[43] Hund A, Ruta N, Liedgens M. Rooting depth and water use efficiency of tropical maize inbred lines, differing in drought tolerance. Plant and Soil. 2009;**318**:311-325. DOI: 10.1007/s11104-008-9843-6

[44] Miguel MA, Postma JA, Lynch JP. Phene synergism between root hair length and basal root growth angle for phosphorus acquisition. Plant Physiology. 2015;**167**:1430-1439. DOI: 10.1104/pp.15.00145

[45] Lopez-Arredondo DL, Leyva-Gonzalez MA, Gonzalez-Morales SI, Lopez-Bucio J, Herrera-Estrella L. Phosphate nutrition: improving low-phosphate tolerance in crops phosphate nutrition. Annual Review of Plant Biology. 2014;**65**:95-123. DOI: 10.1146/arplant-050213-035949

[46] Dong B, Ryan PR, Rengel Z, Delhaize E. Phosphate uptake in Arabidopsis thaliana: dependence of uptake on the expression of transporter genes and internal phosphate concentrations. Plant, Cell & Environment. 1999;**22**:1455-1461

[47] Stigter KA, Plaxton WC. Molecular mechanisms of phosphorus metabolism and transport during leaf senescence. Plants. 2015;**4**:773-798. DOI: 10.3390/plants4040773

[48] Gregersen PL. Senescence and nutrient remobilization in crop plants. In: First PB, Hawkesford MJ, editors. The Molecular and Physiological Basis of Nutrient Use Effi ciency in Crops. Hoboken, New Jersey, United States: John Wiley & Sons, Inc; 2011. p. 83

[49] Adem GD, Id YU, Enrico P, Id H, Id MW. Genetic and physiological traits for internal phosphorus utilization efficiency in rice. PLoS One. 2020;**15** (11):e0241842

[50] Kouas S, Debez A, Slatni T, Labidi N, Drevon JJ, Abdelly C. Root proliferation, proton efflux, and acid phosphatase activity in common root proliferation, proton efflux, and acid phosphatase activity in common bean (*Phaseolus vulgaris*) under phosphorus shortage. The Journal of Plant Biology. 2009;**52**: 395-402. DOI: 10.1007/s12374-009-9050-x

[51] Lopez-Cantarero I, Lorente FA, Romero L. Are chlorophylls good indicators of nitrogen and phosphorus levels? Journal of Plant Nutrition. 1994; **17**(6):979-990. DOI: 10.1080/01904169409364782

[52] Qiu J, Israel DW. Diurnal starch accumulation and utilization in phosphorus-deficient soybean plants 1. Plant Physiology. 1992;**98**:316-323. DOI: 10.1104/pp.98.1.316

[53] Blakeley SD, Dennis ANDDT. Molecular approaches to the manipulation of carbon allocation in plants. Canadian Journal of Botany. 1993; **71**:765-775

[54] Balemi T, Negisho K. Management of soil phosphorus and plant adaptation mechanisms to phosphorus stress for sustainable crop production: A review. Journal of Soil Science and Plant Nutrition. 2012;**12**(3):547-561

[55] Shenoy VV, Kalagudi GM. Enhancing plant phosphorus use efficiency for sustainable cropping. Biotechnology Advances. 2005;**23**:501-513. DOI: 10.1016/j.biotechadv.2005.01.004

[56] Simpson RJ et al. Strategies and agronomic interventions to improve the 56, 77, 81, 82phosphorus-use efficiency of farming systems. Plant and Soil. 2011; **349**(1):89-120. DOI: 10.1007/s11104-011-0880-1

[57] Wang L, Shen J. Root/rhizosphere management for improving phosphorus use efficiency and crop productivity. Better Crops. 2019;**13**(1):36-39. DOI: 10.24047/BC103136

[58] Johnston AM, Bruulsema TW. 4R nutrient stewardship for improved nutrient use efficiency. Procedia Engineering. 2014;**83**:365-370. DOI: 10.1016/j.proeng.2014.09.029

[59] (The International Plant Nutrition Institute) IPNI. 4R Plant Nutrition: A Manual for Improving the Management of Plant Nutrition. Vol. 1. GA, USA: Spring, E. Norcross; 2012. p. 5519

[60] Iqbal Z, Latif A, Ali S, Iqbal MM. Effect of fertigated phosphorus on P use efficiency and yield of wheat and maize. Songklanakarin Journal of Science and Technology (SJST). 2003;**25**(697):702

[61] Rahim A, Ranjha AM, Waraich EA. Effect of phosphorus application and irrigation scheduling on wheat yield and phosphorus use efficiency. Soil and Environment. 2010;**29**(1):15-22

[62] Mclaughlin MJ, Mcbeath TM, Smernik R, Stacey SP, Ajiboye B, Guppy C. The chemical nature of P accumulation in agricultural soils—Implications for fertiliser management

and design: An Australian perspective. Plant and Soil. 2011;**349**:69-87. DOI: 10.1007/s11104-011-0907-7

[63] Jing J, Rui Y, Zhang F, Rengel Z, Shen J. Localized application of phosphorus and ammonium improves growth of maize seedlings by stimulating root proliferation and rhizosphere acidification. Field Crops Research. 2010;**119**:355-364. DOI: 10.1016/j.fcr.2010.08.005

[64] Ma Z et al. pH-responsive controlled-release fertilizer with water retention via atom transfer radical polymerization of acrylic acid on mussel-inspired initiator. Journal of Agricultural and Food Chemistry. 2013;**61**:5474-5482

[65] Rehim A, Farooq M, Ahmad F, Hussain M. Band placement of phosphorus improves the phosphorus use efficiency and wheat productivity under different irrigation regimes. International Journal of Agriculture and Biology. 2012;**14**(5):727-733

[66] Marschner H. Mineral Nutrition of Higher Plants. 3rd ed. San Diego: Elsevier; 2012

[67] Shen J, Yuan L, Zhang J, Li H, Bai Z, Chen X. Phosphorus dynamics: From soil to plant 1. Plant Physiology. 2011;**156**: 997-1005. DOI: 10.1104/pp.111.175232

[68] Witt C, Dobermann A. A site-specific nutrient management approach for irrigated, lowland rice in Asia. Better Crop International. 2002;**16**(1):20-24

[69] Cassman KG, Dobermann A, Walters DT. Agroecosystems, nitrogen-use efficiency, and nitrogen management. Agronomy and Horticulture. 2002;**31**(2):132-140

[70] Hinsinger P. Bioavailability of soil inorganic P in the rhizosphere as affected by root-induced chemical changes: A review. Plant and Soil. 2001; **237**:173-195. DOI: 10.1023/A

[71] Banik SSG, Midya PA, Sarkar BK. Wheat and chickpea intercropping systems in additive series experiment: Advantages and weed smothering. European Journal of Agronomy. 2006; **24**:325-332

[72] Dhima KV, Lithourgidis AS, Vasilakoglou IB, Dordas CA. Competition indices of common vetch and cereal intercrops in two seeding ratio. Field Crops Research. 2007;**100**:249-256

[73] Javanmard A, Dabbagh A, Nasab M, Javanshir A, Moghaddam M, Janmohammadi H. Forage yield and quality in intercropping of maize with different legumes as double-cropped forage yield and quality in intercropping of maize with different legumes as double-cropped. Journal of Food, Agriculture and Environment. 2009;7 (1):163-166

[74] Rojas-downing MM, Nejadhashemi AP, Harrigan T, Woznicki SA. Climate risk management climate change and livestock: Impacts, adaptation, and mitigation. Climate Risk Management. 2017;**16**:1-19. DOI: 10.1016/j.crm.2017.02.001

[75] Zaeem M et al. The potential of corn-soybean intercropping to improve the soil health status and biomass production in cool climate boreal ecosystems. Scientific Reports. 2019;**9**(13148):1-17. DOI: 10.1038/s41598-019-49558-3

[76] Sun B, Gao Y, Wu X, Ma H. The relative contributions of pH, organic anions, and phosphatase to rhizosphere soil phosphorus mobilization and crop phosphorus uptake in maize/alfalfa polyculture. Plant and Soil. 2019;**447** (1-2):117-133. DOI: 10.1007/s11104-019-04110-0

[77] Hinsinger P, Betencourt E, Bernard L, Brauman A, Plassard C, Shen J. Soil phosphorus acquisition in the rhizosphere of intercropped species 1. Plant Physiology. 2011;**156**:1078-1086

[78] Zhang Q et al. Effects of six-year biochar amendment on soil aggregation, crop growth, and nitrogen and phosphorus use ef fi ciencies in a rice-wheat rotation. Journal of Cleaner Production. 2020;**242**:118435

[79] Li C, Dong Y, Li H, Shen J, Zhang F. Shift from complementarity to facilitation on P uptake by intercropped wheat neighboring with faba bean when available soil P is depleted. Scientific Reports. 2016; **18663**:1-9. DOI: 10.1038/srep18663

[80] Wang L, Hou B, Zhang D, Li H, Rengel Z, Shen J. The niche complementarity driven by rhizosphere interactions enhances phosphorus-use efficiency in maize/alfalfa mixture. Food and Energy Security. 2020;**9**(e252):1-14. DOI: 10.1002/fes3.252

[81] Bücking H, Liepold E, Ambilwade P. The role of the mycorrhizal symbiosis in nutrient uptake of plants and the regulatory mechanisms underlying these transport processes. In: Dhal NK, Sahu SC Editors. London: Intechopen. Plant Science. 2012:107-138. ISBN: 978-953-51-0905-1

[82] Qiu B, Wang Y. Phylogenetic distribution and evolution of mycorrhizas in land plants. Mycorrhiza. 2006;**16**:299-363. DOI: 10.1007/s00572-005-0033-6

[83] Smith FA, Smith SE. What is the significance of the arbuscular mycorrhizal colonisation of many economically important crop plants? Plant and Soil. 2011;**348**:63-79

[84] Soka G, Ritchie M. Arbuscular mycorrhizal symbiosis and ecosystem processes: Prospects for future research in tropical soils. Open Journal of Ecology. 2014;**4**(1):11-22

[85] Thingstrup I, Kahiluoto H, Jakobsen I. Arbuscular mycorrhizal fungi at two levels of P fertilization phosphate transport by hyphae of field communities of arbuscular mycorrhizal fungi at two levels of P fertilization. Plant and Soil. 2000;**221**:181-187. DOI: 10.1023/A

[86] Casierra-Posada F, Peña-Olmos J, Peñaloza J, Roveda G. Influence of shading and mycorrhizae on growth of Lulo plants (*Solanum quitoense* lam.). Revista UDCA Actualidad & Divulgación Científica. 2013;**16**(1):61-70

[87] González O, Walter O. Determination of mycorrhizal dependency of Lulo. Acta Biológica Colombiana. 2008;**13**(2):163-114

[88] Bolan NS. A critical review on the role of mycorrhizal fungi in the uptake of phosphorus by plants. Plant and Soil. 1991;**134**:189-207

[89] Campos P, Borie F, Cornejo P, López-ráez JA, López-garcía Á, Seguel A. Phosphorus acquisition efficiency related to root traits: Is mycorrhizal Symbiosis a key factor to wheat and barley cropping? Frontiers in Plant Science. 2018;**9**(752):1-21. DOI: 10.3389/fpls.2018.00752

[90] Igual JM, Valverde A, Cervantes E, Velázquez E. Review article phosphate-solubilizing bacteria as inoculants for agriculture: Use of updated molecular techniques in their study. Agronomie. 2001;**21**:561-568. DOI: 10.1051/agro

[91] Khan MS, Ahmad E, Zaidi A, Oves M. Functional Aspect of Phosphate-Solubilizing Bacteria: Importance in Crop Production. Berlin Heidelberg:

Springer-Verlag; 2013. pp. 237-263. DOI: 10.1007/978-3-642-37241-4

[92] Viruel E, Erazzú LE, Calsina LM, Ferrero MA, Lucca ME, Siñeriz F. Inoculation of maize with phosphate solubilizing bacteria: Effect on plant growth and yield. Journal of Soil Science and Plant Nutrition. 2014;**14**(4):819-831

[93] Alam S, Khalil S, Ayub N, Rashid M. In vitro solubilization of inorganic phosphate by phosphate solubilizing microorganisms from maize rhizosphere. International Journal of Agriculture and Biology. 2002;**4**(4):1-6

[94] Gyaneshwar P, Kumar GN, Parekh LJ, Poole PS. Role of soil microorganisms in improving P nutrition of plants. Plant and Soil. 2002;**245**(1):133-143

[95] Richardson AE et al. Plant and microbial strategies to improve phosphorus efficiency of agriculture plant and microbial strategies to improve the phosphorus efficiency of agriculture. Plant and Soil. 2011;**349**:121-156. DOI: 10.1007/s11104-011-0950-4

[96] Trenkel ME. Slow- and Controlled-Release and Stabilized Fertilizers: An Option for Enhancing Nutrient Efficiency in Agriculture. 2nd ed. Paris, France. 2010. 14-163. ISBN 978-2-9523139-7-1

[97] Liang R, Yuan H, Xi G, Zhou Q. Synthesis of wheat straw-g-poly (acrylic acid) superabsorbent composites and release of urea from it. Carbohydrate Polymers. 2009;**77**(2):181-187

[98] Kordlagharia MP, Rowell DL. The role of gypsum in the reactions of phosphate with soils. Geoderma. 2006; **132**:105-115. DOI: 10.1016/j.geoderma.2005.04.022

[99] Fageria NK, Baligar VC, Li YC. The role of nutrient efficient plants in improving crop yields in the twenty first century. Journal of Plant Nutrition. 2008;**31**(6):1121-1157. DOI: 10.1080/01904160802116068

[100] Takahashi T, Dahlgren RA. Nature, properties and function of aluminum-humus complexes in volcanic soils. Geoderma. 2016;**263**:110-121. DOI: 10.1016/j.geoderma.2015.08.032

[101] Antoniadis V, Hatzis F, Bachtsevanidis D, Koutroubas SD. Phosphorus availability in low-P and acidic soils as affected by liming and P addition. Communications in Soil Science and Plant Analysis. 2015;**2014**:1288-1298

[102] Opala PA. Influence of lime and phosphorus application rates on growth of maize in an acid soil. Advances in Agriculture. 2017;**2017**:1-6

[103] Environ PS, Ochecová P, Mercl F, Košnář Z, Tlustoš P. Fertilization efficiency of wood ash pellets amended by gypsum and superphosphate in the ryegrass growth. Plant, Soil and Environment. 2017;**63**(2):47-54. DOI: 10.17221/142/2016-PSE

[104] Murtaza B, Murtaza G, Sabir M, Owens G, Imran M, Shah GM. Amelioration of saline-sodic soil with gypsum can increase yield and nitrogen use efficiency in rice- wheat cropping system. Archives of Agronomy and Soil Science. 2016;**63**(9):1267-1280. DOI: 10.1080/03650340.2016.1276285

[105] Xiang YAN, Ji-yun JIN, Ping HE, Ming-zao L. Recent advances on the technologies to increase fertilizer use efficiency. Agricultural Sciences in China. 2008;7(4):469-479

[106] Cordell D, White S. For achieving phosphorus security. Agronomy. 2013;**3**: 86-116

[107] Jacobs DF. Variation in nutrient release of introduction. Forest and Conservation Nursery Associations. 2005;**35**:1-6

[108] El Diwani G, Motawie N, Shaarawy HH, Shalaby MS. Nitrogen slow release biodegradable polymer based on oxidized starch prepared via. Journal of Applied Sciences Research. 2013;**9**(3): 1931-1939

[109] Opala P, Othieno CO, Peter K. Effects of combining organic materials with inorganic phosphorus sources on maize yield and financial benefits in western Kenya. Experimental Agriculture. 2010;**46**(1):23-34. DOI: 10.1017/S0014479709990457

[110] Otinga AN et al. Field crops research partial substitution of phosphorus fertiliser by farmyard manure and its localised application increases agronomic efficiency and profitability of maize production. Field Crops Research. 2013;**140**:32-43. DOI: 10.1016/j.fcr.2012.10.003

[111] Noushahi H et al. Improving phosphorus use efficiency by agronomical and genetic means. World Journal of Agricultural Sciences. 2019;**15** (2):47-53. DOI: 10.5829/idosi.wjas.2019.48.54

[112] Arif M, Ilyas M, Riaz M, Ali K, Shah K. Biochar improves phosphorus use efficiency of organic-inorganic fertilizers. Field Crops Research. 2017; **214**:25-37. DOI: 10.1016/j.fcr.2017.08.018

[113] Mitran T, Mani PK. Effect of organic amendments on rice yield trend, phosphorus use efficiency, uptake, and apparent balance in soil under long-term rice- wheat rotation. Journal of Plant Nutrition. 2017;**40**(9):1312-1322. DOI: 10.1080/01904167.2016.1267205

[114] Von Wandruszka R. Phosphorus retention in calcareous soils and the effect of organic matter on its mobility. Geochemical Transactions. 2006; **7**(6):1-8. DOI: 10.1186/1467-4866-7-6

[115] Cong PT, Merckx R. Improving phosphorus availability in two upland soils of Vietnam using Tithonia diversifolia H. Plant and Soil. 2005;**269**: 11-23. DOI: 10.1007/s11104-004-1791-1

[116] Andriamananjara A, Rakotoson T, Razafimbelo T, Rabeharisoa L, Dominique MR. Farmyard manure improves phosphorus use efficiency in weathered P deficient soil. Nutrient Cycling in Agroecosystems. 2019;**3**:1-19. DOI: 10.1007/s10705-019-10022-3

[117] Wahbi S et al. Impact of wheat/Faba bean mixed cropping or rotation systems on soil microbial functionalities. Frontiers in Plant Science. 2016;**7**:1364. DOI: 10.3389/fpls.2016.01364

Chapter 2

Effect of Balanced and Integrated Crop Nutrition on Sustainable Crop Production in a Classical Long-Term Trial

Melkamu Jate and Joachim Lammel

Abstract

The classical long-term trial at Hanninghof was established in 1958 on loamy sand soil in Duelmen, Germany to study the long-term effects of different nutrient management strategies. The impact of balanced mineral fertilizer application and integrating farmyard manure (FYM) with mineral fertilizer on indicators of sustainable crop production are evaluated in comparison to unbalanced nutrition. Crop rotation since 1958 was potato, followed by winter rye and oat. After 2008, the rotation was silage maize, winter rye, and potato to adjust the trial to current farm practice, but the treatments remained the same: a control plot without fertilizer; FYM alone; and mineral P + K, N, N + P, N + K, N + P + K, and N + P + K + Mg fertilizers with and without FYM. The effect of each treatment on crop yield, revenue, sustainable yield index, water and nutrient use efficiencies, soil nutrient and carbon contents, and soil pH are presented. Evaluation of the 62 years data shows that unbalanced nutrition caused by omitting nutrients and application of only FYM as organic nutrition reduced crop yield and revenue, led to inefficient use of resources and nutrients, and a depletion of soil fertility with negative implications on sustainability. Application of mineral fertilizer N + P + K + Mg as the balanced nutrition and supplementing FYM with mineral fertilizer as the integrated nutrition had social, economic and environmental benefits indicating sustainable crop production.

Keywords: balanced nutrition, integrated nutrition, soil fertility, sustainable crop production

1. Introduction

Long term trials (LTTs) are conducted on a stationary site for many years and classified as Young, Medium, and Classical, respectively in age less than 20, 20–50, and older than 50 years [1, 2]. They are appropriate to study the sustainability of crop production which is defined as the ability to produce the required crop yield and quality to satisfy present and future food demand, while protecting the environment. Population and economic growths are estimated to result in a 50% increase in the demand for food by 2050 with little scope to expand the agricultural area [3]. Thus, a sustainable increase of crop yield per area is required to meet the rising demand for food.

This target requires improvement of yield through integration of productive crop varieties, fertile soils, adequate water supply, sufficient plant nutrients with efficient use, protection of crops against weeds, diseases and pests, and post-harvest care [4]. Continuous crop yield increases are mostly determined by the improvement of crop varieties. Improved crop varieties require advanced cultivation practices, best nutrient application strategies, and pre- and post-harvest crop protection [5]. Improvement of crop nutrition is one of the essential management factors to increase yield.

A trial conducted in India for example, showed that improved cultivars along with balanced nutrition resulted the highest yield increase in range of 92 to 204% over the farmer's practice [6]. Vyn (2014) said "global maize yields will not be able to continually boost to achieve food security without providing adequate and balanced nutrients" [7]. The synergy between improved genetic and adequate nutrient supply sustained the increased production of rice and wheat for nearly three decades in India; however, in recent years the high productivity is stagnating or declining in spite of supplying increased N, P and K fertilizer rates, because of unbalanced nutrient application [8].

Soil fertility is the major environmental factor and is viewed as the capacity of the soil to retain, cycle, and supply essential nutrients to support crop growth for a long time [9]. The relationship of nutrient application and soil fertility is reliably studied in the LTTs, because soil fertility develops gradually and therefore, evaluation of its effect on crop production requires monitoring over a long time and a proper documentation of data [10]. The LTTs are the right tool to study changes that can take decades before they become visible, for example: trends of crop yield and effects of the environment on agriculture or vice versa [11]. Since agriculture is removing nutrients from the soil an efficient replacement of nutrients back into the soil is required to sustain crop yields [12]. The target to increase crop yield per area requires avoiding nutrient mining, maintaining soil fertility, and minimizing nutrient loss to protect the environment.

Balanced nutrition is the key to sustainable crop production and maintenance of soil health with both economic and environmental benefits. An unbalanced nutrition results in a low nutrient use efficiency, poor economic returns, and high environmental pollution [13]. The Law of the Minimum states: "If one of the essential growth factors/nutrient is deficient, plant growth will be limited even when all other factors/nutrient are sufficiently available that growth is improved by the application of deficient factor/nutrient". Dev (1998) viewed the balanced nutrition as "a best management practice refer to the application of essential nutrients in optimum quantity and proportion including proper application methods and time for the specific soil, crop, and climate conditions" [14]. It ensures accessibility of crops to an adequate nutrient supply at every growth stage to avoid over or under-supply enabling the crop for a strong, healthy and productive growth while minimizing pollution of the environment [15]. It can be further defined as the application of right sources of nutrients in an adequate amount and ratio with optimum methods at the time required to support healthy crop growth to increase yield and quality.

Integrated crop nutrition is the combined application of organic and mineral fertilizers to increase yield and to improve soil fertility. Organic fertilizer alone can often not fully satisfy the nutritional demand of crops, because it contains inadequate and unbalanced nutrients [16]. It may not be available in sufficient quantities. Supplementing organic fertilizer with mineral fertilizer is needed to improve nutrient availability and increase crop yield.

The Hanninghof classical LTT was established in 1958. Three strategies of crop nutrition are compared: (1) Balanced nutrition, (2) Integrated nutrition, the combination of farmyard manure (FYM) with mineral fertilizer, and (3) Organic nutrition, application of FYM only. The effects of these different treatments on crop yield, economic revenue, sustainable yield index, water and nutrient use efficiencies and soil nutrient content, organic matter and pH are measured to evaluate social, economic, and environmental benefits of best strategies of nutrient application. The objective of the trial is to study the long-term impacts of different nutrient management on parameters of crop productivity and soil fertility to quantify sustainable crop production.

2. Materials and methods

2.1 Location and history

The Hanninghof LTT is one of few classical LTT in the world. It is located near Duelmen in Western Germany. Crop rotation started with potato cultivation in 1958, followed by winter rye in 1959 and oat in 1960. The sequence of rotation changed to silage maize, winter rye, and potato after 2008 to adjust the trial to current agricultural practices, but the basic setup of the trial remains the same. During 1958–2020, potato, winter rye, oat, and silage maize were cultivated respectively 19, 21, 17, and 5 times.

2.2 Soil and climate

The trial was established on a loamy sandy soil with the following initial soil parameters: P_2O_5 13.3 mg (100 g) $^{-1}$, K_2O 10 mg (100 g) $^{-1}$, Mg 2.1 mg (100 g) $^{-1}$, organic carbon 2.1%, N total 0.1%, and pH 5 at soil depth 0–30 cm. The annual rainfall and yearly mean air temperature were, respectively 469–1273 mm and 7.7–12.3°C during 1958–2020.

2.3 Layout

The trial is a two-factorial experiment in a split-plot with a randomized complete block design. The cultivated area of the trial is 0.3 ha (72 m × 42 m). The field is split into two parts, one receiving FYM every 3 years during 1958–2008 and yearly since 2009 and the other part is without FYM. Each of the two parts is subdivided into 32 plots, i.e., 64 plots in total. The gross area of a plot is 4.5 m × 10.5 m with a harvested area of 3.5 m × 9.5 m to avoid the border effect.

2.4 Treatment

Sixteen treatments were established as shown in **Table 1**. Each treatment is replicated four times and randomly assigned to plots. In 1960, a treatment with N only (#8 and #16) was established. Since the trial was already ongoing for 2 years, new control treatments were assigned. The new control treatments (#7 and #15) are omitted from data evaluation (**Table 1**), because they were not different from treatments #2 and #10.

FYM plus mineral fertilizer		Mineral fertilizer without FYM	
#	Treatments	#	Treatments
1	FYM + N + P	9	N + P
2	FYM without mineral fertilizers	10	Control (without mineral fertilizers)
3	FYM + N + K	11	N + K
4	FYM + N + P + K	12	N + P + K
5	FYM + P + K	13	P + K
6	FYM + N + P + K+ Mg	14	N + P + K+ Mg
7	FYM without mineral fertilizers	15	Control (without mineral fertilizers)
8	FYM + N	16	N

Table 1.
Description of treatments.

2.5 Nutrient application

Mineral fertilizer nitrogen (N), phosphorous (P), potassium (K), and magnesium (Mg) rates were the same for the treatments with and without FYM during 1958–2008. FYM was applied as pig manure at a rate of 25 t ha^{-1} once every 3 years in spring. Nutrient content of FYM is calculated based on 7 kg N-total, 6.7 kg P_2O_5, 7.2 kg K_2O and 2.2 kg MgO per ton of pig manure [17].

After 2008, the trial was adjusted to reflect recent crop rotation and nutrient application. Oat was replaced by silage maize. FYM was replaced by cattle slurry and applied annually at the rate of 30, 20, and 20 m^3 ha^{-1}, respectively during silage maize, winter rye, and potato cultivation. The nutrient content of FYM was considered in the total nutrient application rate to make the nutrient input with FYM and without FYM comparable. The nutrient content of cattle slurry was analyzed every year in the laboratory. Nutrient rates for potato, winter rye, oat and silage maize are given in **Table 2**.

N, P, K, and Mg from mineral fertilizers were applied as calcium ammonium nitrate (CAN) with 4% MgO, triple supper phosphate, potassium chloride, and magnesium nitrate respectively. Since 2013, N was applied as CAN with 6% S to avoid 4% MgO content of CAN that resulted in a reduction of the treatment effect of Mg on crop yield. Since 1958, lime (CaO) was applied to the whole field at a rate of 1000 kg ha^{-1} every 3 years to stabilize soil pH. Since 2009, S fertilizer was applied every year at a rate of 20 kg S per ha on the whole field to avoid S deficiency. Pig manure was applied 10 days before potato planting during 1958–2008. Since 2009, cattle slurry was applied 10 days before silage maize and potato planting, and at the early vegetative stage of winter rye. Mineral fertilizer N was applied once at planting for potato. It was split applied for winter rye at early vegetative, stem elongation, and booting; for oat at seeding and booting; and for maize at seeding and early vegetative growth stages. P, K and Mg mineral fertilizers were applied once at the planting of potato, oat, and silage maize; and at the early vegetative stage of winter rye cultivation.

2.6 Analysis of crop and soil parameters

Crop fresh and dry matter yields were recorded. The crop samples were dried in a drying cabinet at 70°C. Soil samples were collected before crop seeding (planting) at

Crop	Years	FYM + mineral fertilizer (kg ha $^{-1}$)				Mineral fertilizer alone (kg ha $^{-1}$)			
		N	P_2O_5	K_2O	MgO	N	P_2O_5	K_2O	MgO
Potato	1958–1979	175 + 100	168 + 90	180 + 160	55 + 50	100	90	160	50
	1979–2006	175 + 140	168 + 90	180 + 160	55 + 50	140	90	160	50
	2013–2018	68 + 88	29 + 27	83 + 77	18 + 10	140	60	160	30
Rye	1959–1980	0 + 60	0 + 90	0 + 120	0 + 50	60	90	120	50
	1980–2007	0 + 140	0 + 90	0 + 120	0 + 50	140	90	120	50
	2010–2018	77 + 100	31 + 31	85 + 41	20 + 11	150	60	120	30
Oat	1960–2008	0 + 100	0 + 90	0 + 120	0 + 50	100	90	120	50
Maize	2009–2016	132 + 87	55 + 20	135 + 20	35 + 9	170	60	150	38
	2019–2020	78 + 125	35 + 27	95 + 104	24 + 14	200	75	230	44

Table 2.
Nutrient application rate per year during 1958–2020.

0–30 cm soil depth from all 4 plots of each treatment and mixed thoroughly to obtain a uniform sample. Macro and micro nutrient concentrations in the tuber, grain, straw and silage of crop and soil nutrient content, organic matter and pH were analyzed as follows:

N content of crop: Crop dry matter was digested with sulfuric acid and catalyst tablet to produce 50 ml of filtered samples. The N concentration of the sample was determined by continuous flow analysis based on standard operation procedures according to the Kjeldahl method.

Macro and micro nutrients content of crop: The dried crop samples were digested with nitric acid by direct heating in the microwave. The macro and micro nutrient in the digested samples were determined on the ICP-OES (inductively coupled plasma-optical emission spectrometry) according to standard operation procedures.

Soil P and K content: The air-dry soil samples were sieved via 2 mm sieve and mixed with 100 ml calcium acetate and lactate solutions and shaken on the flat shaker for 90 minutes. The plant available P and K contents of filtrate of the soil samples were determined by ICP.

Soil organic matter: The total organic carbon (TOC) was determined by Vario Select Elementary device. The TOC content was calculated from the integral values of the measurement peaks and the calibration coefficients.

Soil pH: The air-dried soil samples were sieved on 2 mm sieve and pH was measured in a 0.01molar $CaCl_2$ solution after 1 hour by pH electrode.

2.7 Data organization and evaluation

Crop yield data were converted to cereal units to aggregate data of different crops along 62 years. The potato tuber, winter rye grain, oat grain, and maize silage yield were multiplied by respectively 0.22, 1.01, 0.85, and 0.18 to convert into cereal unit [18]. The significance differences between average crop yield of treatments were analyzed statistically. The yield data were grouped into 12 periods (1958–1963, 1963–1968, 1968–1973, 1973–1978, 1978–1983, 1983–1988, 1988–1993, 1993–1998, 1998–2003, 2003–2008, 2008–2014, and 2014–2020) to evaluate the trend of crop yield, because crop varieties remained unchanged during 5- or 6-years interval per each period with similar effect on yield.

Crop yields data (1958–2020) were converted to revenue (economic yield) by multiplying annual yields with historical crop prices [19]. The cost of mineral fertilizer was obtained by multiplying the mineral fertilizer rate with historical prices [20]. FYM was regarded free of cost. The economic evaluation included mineral fertilizers cost only, because all other costs of crop production were considered equal for all treatments. Economic benefit (USDha^{-1}) = crop revenue - mineral fertilizer cost.

Sustainable yield index (SYI) was calculated according to Singh et al. (1990) based on the standard deviation of mean to evaluate the stability of yield [21, 22]. SYI = average yield (AY) of treatments minus standard deviation (SD) divided by maximum yield (MY) in different years and treatments.

Green water use efficiency (WUE) was calculated according to Sharma et al. (2013) based on historical rainfall data recorded at the LTT site [23]. WUE (kg yield per mm rainwater) = Yield (kgha^{-1}) divided by cumulative rainfall (mm) from sowing to harvest.

Nutrient use efficiency was calculated according to partial factor productivity [24]. N use efficiency (%) = N removal with N fertilized crop divided by N fertilizer rate and multiplied by 100. The calculation was done similarly for P and K fertilizer use efficiencies.

The soil fertility is measured by nutrient content, organic matter and pH. The soil parameters were organized with a three-year moving average. The changes in soil fertility were evaluated in comparison to the control treatments and initial values measured in 1958.

3. Results

3.1 Effect of balanced and unbalanced nutrition on crop production and soil fertility

3.1.1 Average agronomic and economic yields

The balanced nutrition (N + P + K + Mg treatment) resulted in the highest average cereal unit yield of 5.4 t and an economic benefit of 1216 \$ha^{-1} (**Figure 1**). The average yield of 5.4 tha^{-1} is low because of the low-yielding varieties at the early decades of the trial and a low water holding capacity of the sandy soil at the site. Without fertilizer application, the yield was only 1.9 t and 469 \$ha^{-1}. The application of P and K without N showed almost no increase in crop yield, 2.1 t and 404 \$ha^{-1}. N fertilizer application but omitting P, K, and Mg resulted in a yield of 3.8 t and 831 \$ha^{-1}. The yield declines due to omission of K + Mg, P + Mg, and Mg were respectively 18%, 9%, and 7% and the corresponding income loss were 315\$, 70\$ and 89\$ (**Figure 1**). Omitting K fertilizer leads to a higher yield reduction than omitting P fertilizer because of decreasing K supply from the soil (**Figure 2**). The yield and income loss due to the omission of P were rather small because of the high P content of the soil (**Figure 3**). Application of CAN with 4% MgO during 1958–2013 as a source of N resulted in low effect of omitting Mg fertilizer on crop yield. Application of only FYM decreased yield by 38% and 275\$ha^{-1} (**Figure 1**).

3.1.2 Trend of average agronomic and economic yields

Crop varieties improved during 62 years of the trial which can be seen in the yield increase over time in almost all treatments. The balanced nutrition (N + P + K + Mg treatment) resulted in the highest yield and income compared to nutrient omissions

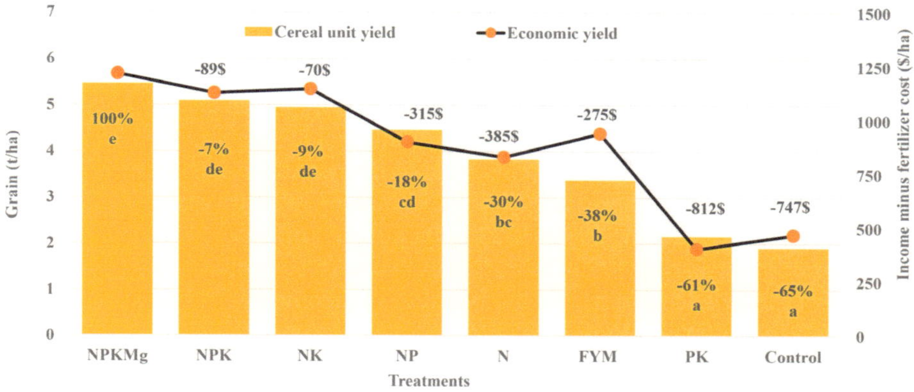

Figure 1.
Effect of balanced nutrition, omitting nutrient, and FYM on average crop yields (n = 62 years); Grain yield with the same letters showed the insignificant difference.

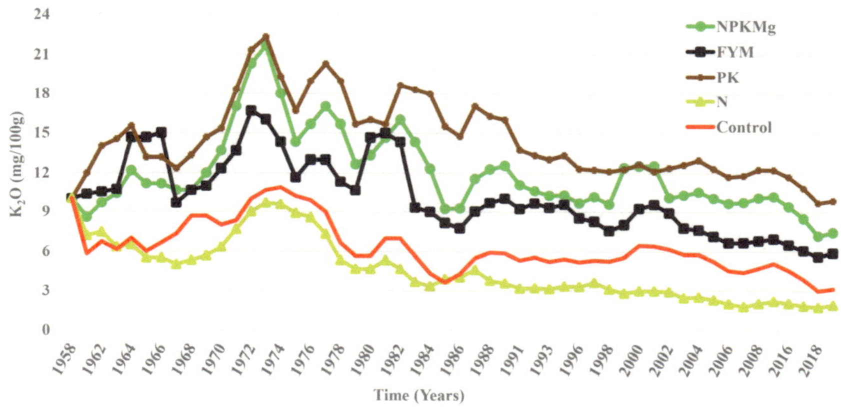

Figure 2.
Effect of balanced nutrition, omitting nutrient, and FYM on soil K content in 0–30 cm depth.

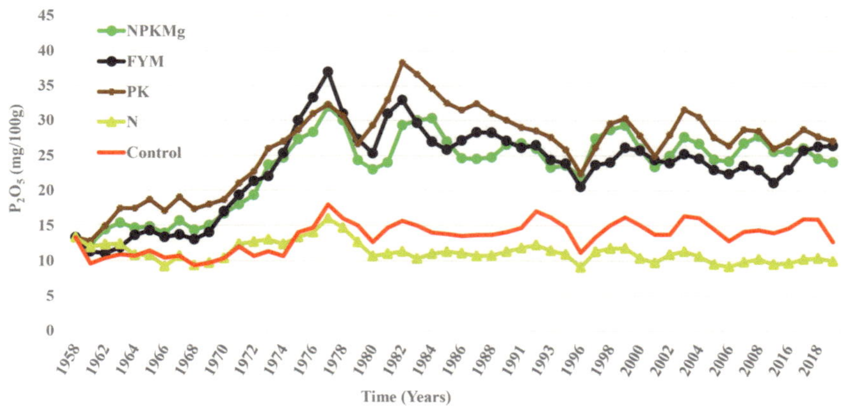

Figure 3.
Effect of balanced nutrition, omitting nutrient, and FYM on soil P content in 0–30 cm depth.

or FYM application without mineral fertilizer (**Figures 4** and **5**). Before 1980, the N fertilizer rate was insufficient to provide the N demand of potato and winter rye. The low N rate of 100 kgha^{-1} for potato and 60 kgha^{-1} for rye cultivation during1958–1980 (**Table 2**) and reduction of potato yield by nematodes infection in 1973–1982 resulted in decreasing yields during 1968–1980. Since 1980, increasing N rate and cultivating nematode resistance potato reversed the trend of decreasing yields. Improvement of crop variety resulted in increasing yields and income in all treatments, however declining cereal prices during 1990–2003 (data not shown) resulted in decreasing crop economic yield during 1993–2003. Compared to the income at the initial phase (1958–1963), the balanced nutrition increased crop income by 1981 \$ha^{-1} at the final (2014–2020) time interval (**Figure 5**).

3.1.3 Sustainable yield index (SYI) and green water use efficiency (WUE)

SYI indicates the stability of crop yields in the long run. The high index shows the low variation of yield increase over years. Application of mineral fertilizers N + P + K + Mg increased SYI and WUE of crop. Omitting nutrient and application

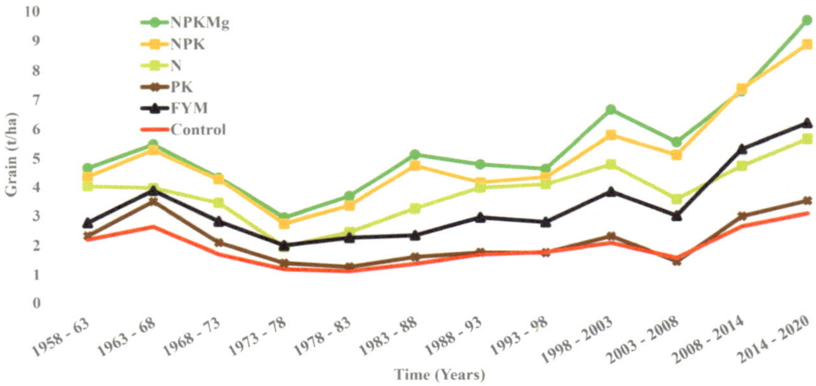

Figure 4.
Effect of balanced nutrition, omitting nutrient, and FYM on the trend of average yield.

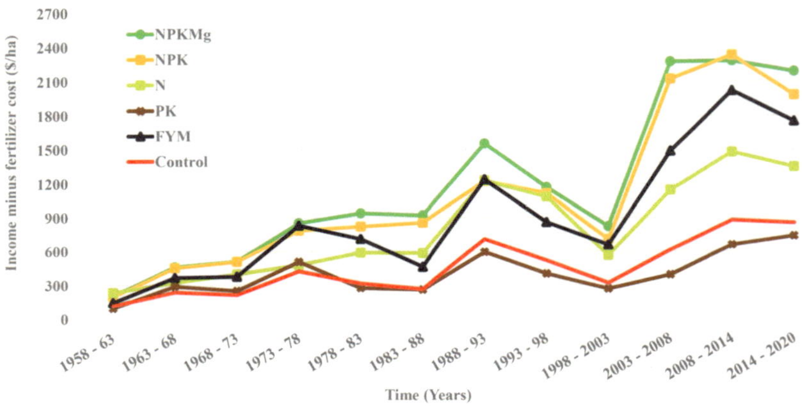

Figure 5.
Effect of balanced nutrition, omitting nutrient, and FYM on the trend of average economic yield.

of only FYM decreased SYI and WUE compared to the balanced nutrition. The WUE was reduced by 63%, 34%, and 7–59%, respectively at without any fertilizer, application of only FYM, and omitting nutrients compared to the N + P + K + Mg treatment (**Figure 6**). The reduction of WUE is directly related to the decline in crop yield because of nutrient omission and application of only FYM.

3.1.4 Nutrient use efficiency

Nitrogen use efficiency (NUE) at the balanced nutrition (N + P + K + Mg) treatment was 77%. Any lack of nutrients reduced the NUE to 49% - 73% (**Figure 7**). P use efficiency (PUE) at the balanced nutrition was 49%. Omitting Mg, K + Mg, and N + Mg resulted, respectively 46%, 42%, and 25% PUE compared to the N + P + K + Mg treatment (**Figure 7**). The balanced nutrition resulted in the highest K use efficiency (KUE) of 84% compared to omitting nutrients and application of only FYM. The KUE at omitting Mg, P + Mg, and N + Mg fertilizers and only FYM

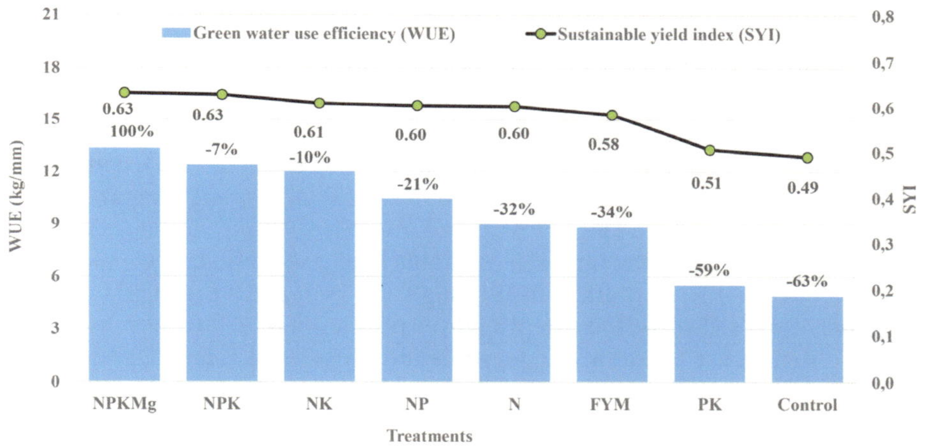

Figure 6.
Effect of balanced nutrition, omitting nutrient, and FYM on average WUE and SYI of the crop (n = 62 years).

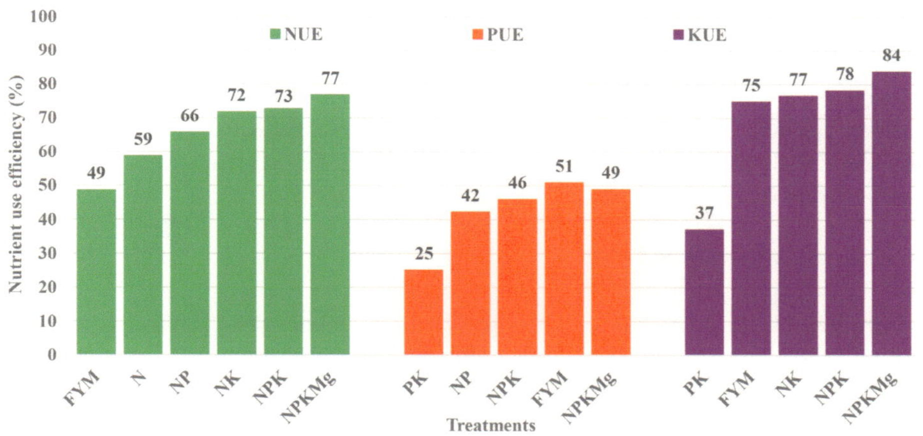

Figure 7.
Effect of balanced nutrition, omitting nutrient, and FYM alone on average N, P, and K use efficiency (n = 62 years).

application were, respectively 78%, 77%, 37%, and 75% (**Figure 7**). The highest nutrient use efficiencies were achieved in the treatment with the balanced supply of nutrients for crop demand. Application of FYM without mineral fertilizer decreased N and K use efficiencies, because of the lower plant availability of N and K in the FYM. P and K fertilizers application without N resulted in a very low PUE and KUE, because of the very low yield and poor recoveries of P and K.

3.1.5 Soil P_2O_5 and K_2O contents

P and K fertilizers application affects soil P_2O_5 and K_2O content. The balanced nutrition (N + P + K + Mg) increased soil P_2O_5 and maintained the soil K_2O in comparison to the omission of P and K respectively.

The P_2O_5 content of loamy sand soil on arable land is classified as 'very low' (below 3), 'low' (4–9), 'medium' (10–18), 'high' (19–32), and 'very high' (above 33) mg P_2O_5 per 100 g soil at 0–30 cm depth [25]. At the start of the trial the soil P content, as well as the K level were therefore classified as medium. P fertilizer application increased soil P_2O_5 content toward very high during 1958–1983: the inadequate N rate during 1958–1980 (**Table 2**) and the limited potato growth in 1973–1982 caused a low crop yield (**Figure 4**) that resulted in an accumulation of P fertilizer in the soil. Increased crop yield after 1982 due to increased N fertilizer rate and cultivating potato variety resistance to nematodes reduced soil P_2O_5 content during 1984–1996. P fertilizer application generally increased soil P_2O_5 content to 'high' level, but omitting P fertilizer reduced the soil P_2O_5 content compared to the initial measurement in 1958 (**Figure 3**).

The K_2O content of loamy sand soil on arable land is classified as 'very low' (below 3), 'low' (4–9), 'medium' (10–18), 'high' (19–32), and 'very high' (above 33) mg K_2O per 100 g soil at 0–30 cm depth [25]. Application of K fertilizer-maintained soil K_2O content at the 'medium' range, while the omission of K fertilizer decreased soil K_2O content to the 'low' level (**Figure 2**). Application of K fertilizer increased the soil K_2O content during the early decades, because of low K removal from the soil. The decreasing soil K_2O content after 1981 was generally driven by combined effects of increased K removal from the soil with high crop yield and loss of K by leaching on sandy soil.

A low crop yield (**Figure 1**) produced a low PUE and KUE (**Figure 7**), at P + K mineral fertilizers application without N, resulted in the highest soil P_2O_5 and K_2O contents (**Figures 2** and **3**). Soil P_2O_5 and K_2O analysis at 30–90 cm in 1987, 2008, and 2018 showed residual P and K fertilizers movement below 30 cm depth. P + K mineral fertilizers application without N increased the soil P_2O_5 and K_2O contents respectively by 43% and 49% in 30–60 cm and by 48% and 96% in 60–90 cm depth compared to the application of N + P + K mineral fertilizers (data not shown).

3.1.6 Soil organic matter and soil pH

Improvement of soil organic matter positively influences soil fertility through its impact on the chemical, physical, and biological properties of a soil. The soil organic matter was measured as soil C (carbon) content. The soil organic C decreased in comparison to the initial value of 2.1% measured in 1958, because crop residues were removed from the field during 1958–2009. It was slightly increased with mineral fertilizer and FYM application compared to the treatment without any fertilizer (**Figure 8**). During 1959–1973, the soil organic C was not measured.

The soil pH was optimized by lime application to avoid the negative effect of pH on nutrient availability. Lime (CaO) applied every 3 years to the whole field at 1000 kgha^{-1} increased soil pH compared to the initial pH measured in 1958 (**Figure 9**).

3.2 Effect of integrating FYM with mineral fertilizer on crop production and soil fertility

3.2.1 Average agronomic and economic yields

Application of FYM plus mineral fertilizer increased yield and income. The highest yield was measured at 6 tha^{-1} in the treatment of FYM plus NP fertilizer. FYM application without mineral fertilizer as organic nutrition only, decreased yield by 44% (**Figure 10**). Application of FYM + N, FYM + NP, FYM + NK, FYM + NPK, and FYM + NPKMg achieved similar yield levels, but FYM + PK fertilizer significantly decreased yield and income due to inadequate availability of N applied as FYM in the treatment. Integrating FYM with NK fertilizer resulted in the highest income measured at 1433 \$ha^{-1} (**Figure 10**). The economic yield at FYM + NP treatment was

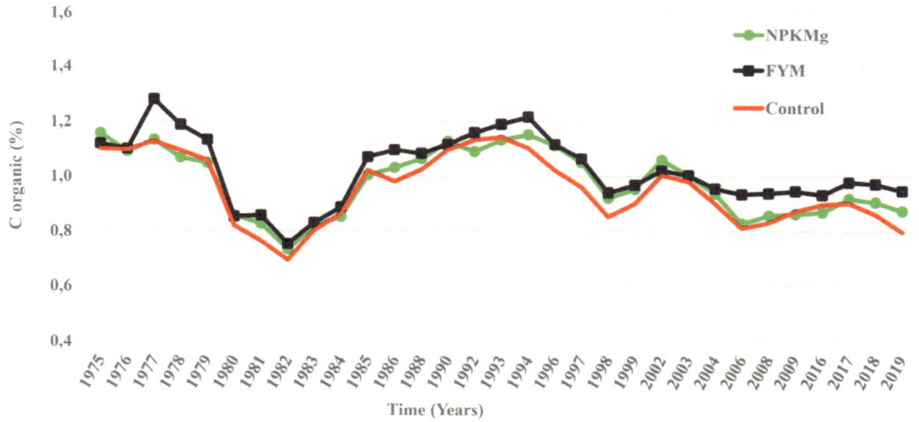

Figure 8.
Effect of balanced nutrition, omitting nutrient, and FYM on soil carbon content in 0–30 cm depth.

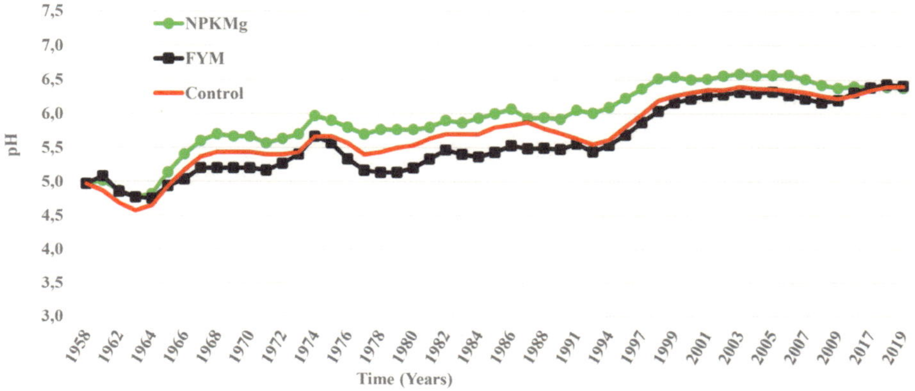

Figure 9.
Effect of balanced nutrition, omitting nutrient, and FYM on soil pH in 0–30 cm depth.

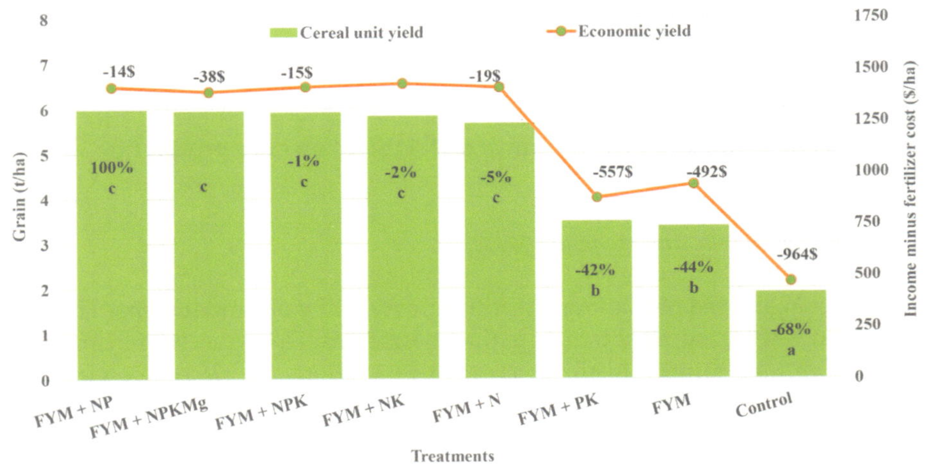

Figure 10.
Effect of integrated nutrition and FYM on average crop yields (n = 62 years); Grain yield with the same letters showed the insignificant difference.

14 $ lower than the economic yield at FYM + NK, because the historical P fertilizer price was higher than K fertilizer price (data not shown). Application of only FYM significantly decreased yield and income, because of the insufficient availability of nutrients in the FYM for the crops.

3.2.2 Trend of average agronomic and economic yields

Improvement of crop varieties during the 62 years of the trial resulted in increasing yield and income trends in all the treatments. The integrated nutrition supported the highest yield and income compared to FYM application only (**Figures 11** and **12**). The decline in yield during 1968–1980 was caused by the low N fertilizer rate in1958–1980 and the reduction of potato growth by nematodes infection in 1973–1982. Increased N fertilizer rate after 1980 and cultivating nematode resistance variety after 1982 reversed decreasing yield levels. The yield was high at integrated nutrition treatment because nutrients were balanced and adequately available compared to only FYM

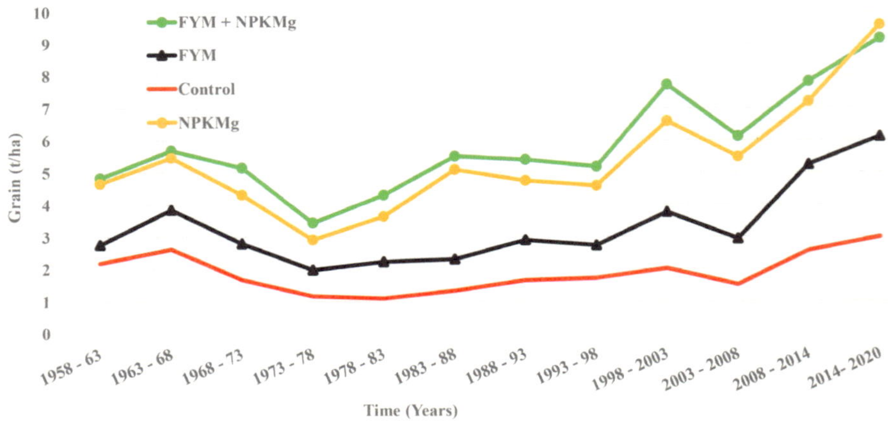

Figure 11.
Effect of integrated nutrition and FYM on the trend of average cereal unit yield.

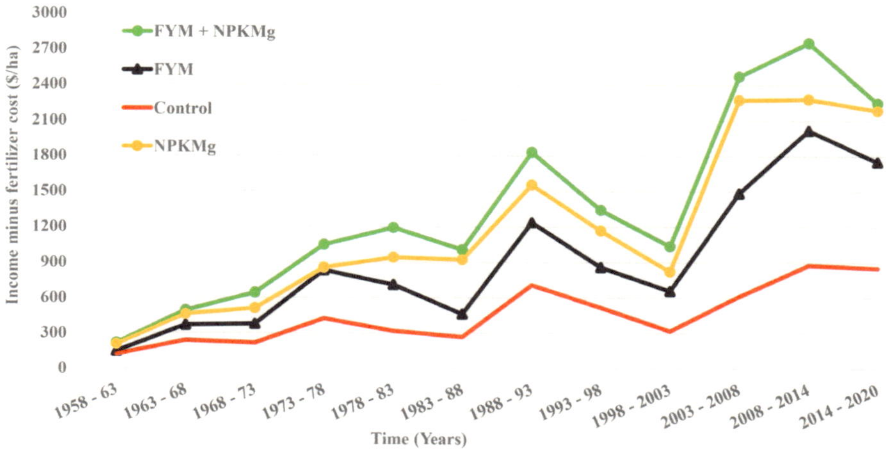

Figure 12.
Effect of integrated nutrition and FYM on the trend of average economic yield.

application. Crop yield and income at integrated nutrition was even higher than at the balanced nutrition (NPKMg) because until 2008, the treatments with FYM received the nutrients from the FYM in addition to nutrients from the mineral fertilizer.

3.2.3 Sustainable yield index (SYI) and green water use efficiency (WUE)

The sustainability of crop production is measured by SYI. A high or low index indicates the level of variations in yield. It is measured as the standard deviations and it is seen as an indicator for sustainability. Nutrient management influences the long-term yield stability. Application of FYM plus mineral fertilizers increased SYI and WUE of the crop (**Figure 13**). WUE decreased by 67% without any fertilizer and by 40% at only FYM compared to the WUE of integrating FYM with NP fertilizer (**Figure 13**). A reduction of crop yield because of nutrient deficiency resulted in a low SYI and inefficient use of water.

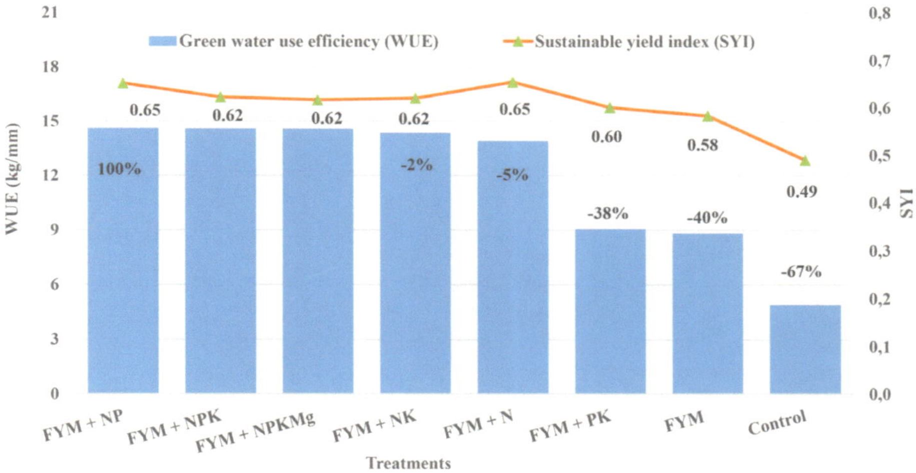

Figure 13.
Effect of integrated nutrition and FYM on average WUE and SYI (n = 62 years).

3.2.4 Nutrient use efficiency

Nitrogen use efficiency (NUE) was 73% at the integrated nutrition treatment (FYM + NP mineral fertilizer). Application of FYM only and FYM plus PK fertilizer decreased NUE to 49% and 51% (**Figure 14**). Low yield (**Figure 10**) due to low N availability in the FYM reduced NUE at the treatments FYM only and FYM plus PK fertilizer. The highest P use efficiency (PUE) of 66% was calculated at FYM + NK fertilizer (**Figure 14**). Low PUE of 51% at the application of only FYM and 32% at FYM plus PK fertilizer were recorded, because inadequate availability of N from FYM reduced yield (**Figure 10**) and P output. Combining FYM with NK or N fertilizer significantly increased PUE. K use efficiency (KUE) was increased to 99% at integrating FYM with NP fertilizer (**Figure 14**). It was decreased to 75% at the application of FYM only and decreased to 47% by omitting mineral nitrogen at FYM plus PK. The nitrogen deficiency in the FYM plus PK fertilizer treatment decreased yield and limited K recovery.

3.2.5 Soil P_2O_5 and K_2O contents

The soil P_2O_5 content indicates the capacity of a soil to supply P for crop growth and it is affected by P fertilizer application. It was increased during 1958–1983, because of inadequate N fertilizer rates during 1958–1980 (**Table 2**) and low yielding potato from 1973 to 1982 resulted in an accumulation of P fertilizer in the soil. Increased crop yield after 1982 due to increased N fertilizer rate and cultivating nematode resistance variety caused a decreasing trend of soil P_2O_5 during 1984–1996 compared to the highest soil P_2O_5 content recorded in1977 and 1982. Integrating FYM with P fertilizer increased soil P_2O_5 content to the 'very high' level. FYM only and FYM plus N or NK fertilizers increased soil P_2O_5 to the 'high' level compared to the initial soil P_2O_5 measured in 1958 (**Figure 15**).

Integrated nutrition and application of FYM only increased soil K_2O during 1958–1980 compared to the initial soil K_2O content (**Figure 16**), it was caused by low crop yield resulting in an accumulation of residual fertilizer K in the soil. The decreasing trend of soil K_2O after 1980 was caused by a combined effect of increased K removal from the soil through high crop yields (**Figure 10**) and K loss by K leaching

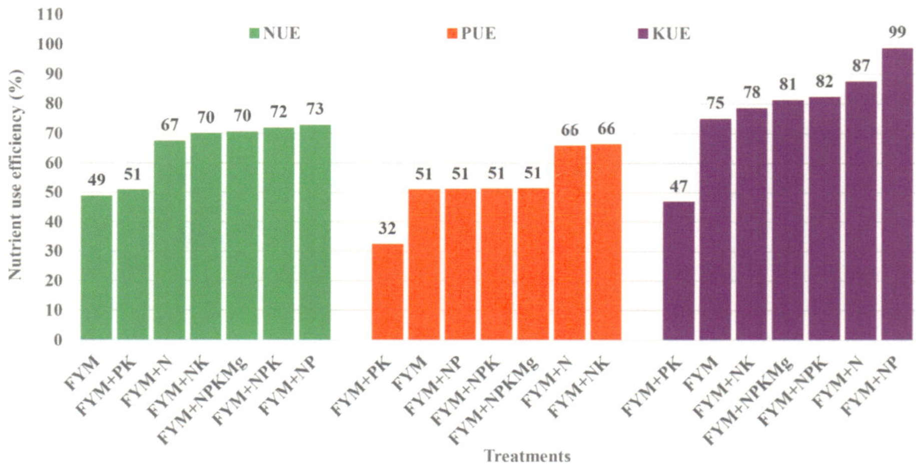

Figure 14.
Effect of integrated nutrition and FYM alone on average N, P, and K use efficiency (n = 62 years).

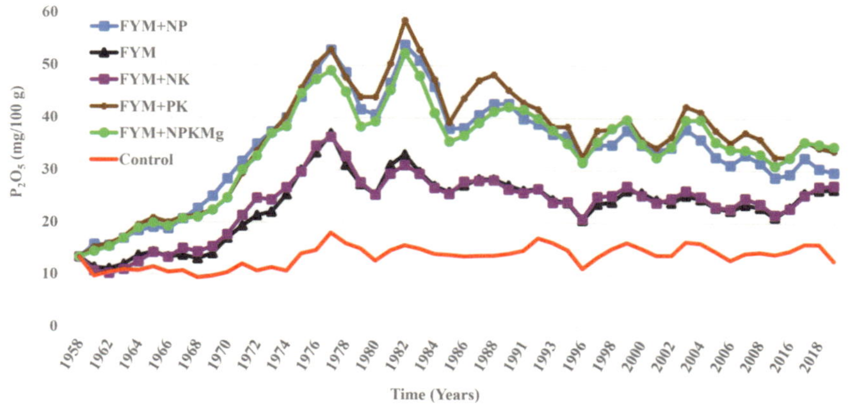

Figure 15.
Effect of integrated nutrition and FYM on soil P content in 0–30 cm depth.

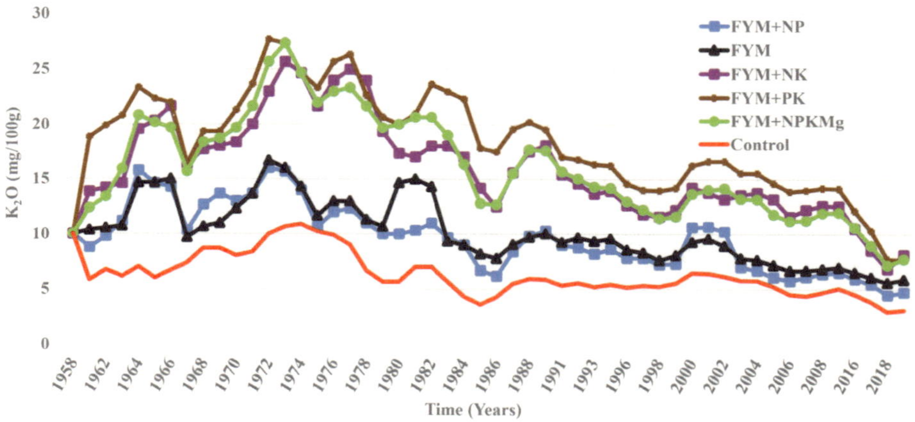

Figure 16.
Effect of integrated nutrition and FYM on soil K content in 0–30 cm depth.

on sandy soil. FYM combined with K fertilizer generally maintained soil K_2O content, but FYM only and FYM plus N or NP fertilizers depleted soil K_2O to the 'low' level after 1982 (**Figure 16**). The negative input–output balance caused a K-mining of the soil of 5 kg K per ha and year at FYM alone, and 41, and 30 kg K per ha per year, respectively at FYM plus N, and FYM plus NP fertilizers and decreased soil K_2O content. FYM plus PK fertilizer resulted in the highest soil K_2O content (**Figure 16**) due to low crop yield (**Figure 10**) and inefficient use of K fertilizer (**Figure 14**). Analysis of soil K_2O in 30–90 cm in 1987, 2008, and 2018 showed residual K fertilizer movement below 30 cm depth. The soil K_2O content increased by 37% in 30–60 cm and 22% in 60–90 cm depth at FYM plus PK fertilizer compared to the FYM plus NPK fertilizer (data not shown).

3.2.6 Soil organic matter and soil pH

The soil organic matter improves soil fertility by its influence on the chemical, physical, and biological properties of a soil. It was measured as an organic fraction of soil C. The soil organic C decreased in comparison to the initial value of 2.1%

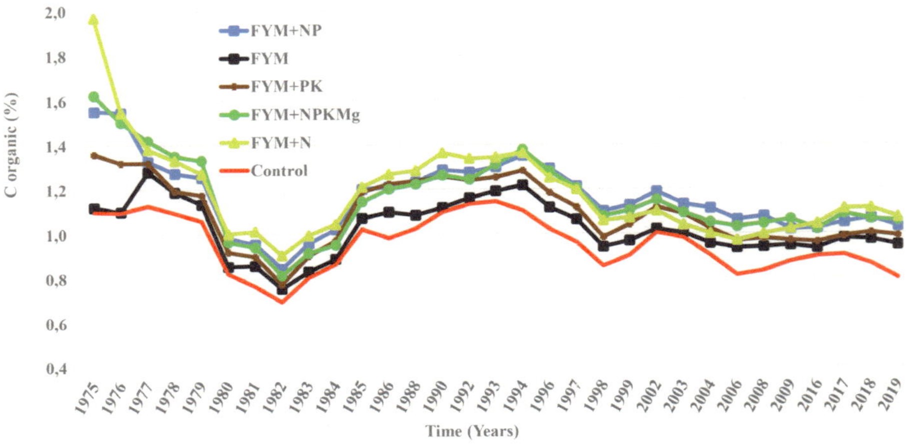

Figure 17.
Effect of integrated nutrition and FYM on soil carbon content in 0–30 cm depth.

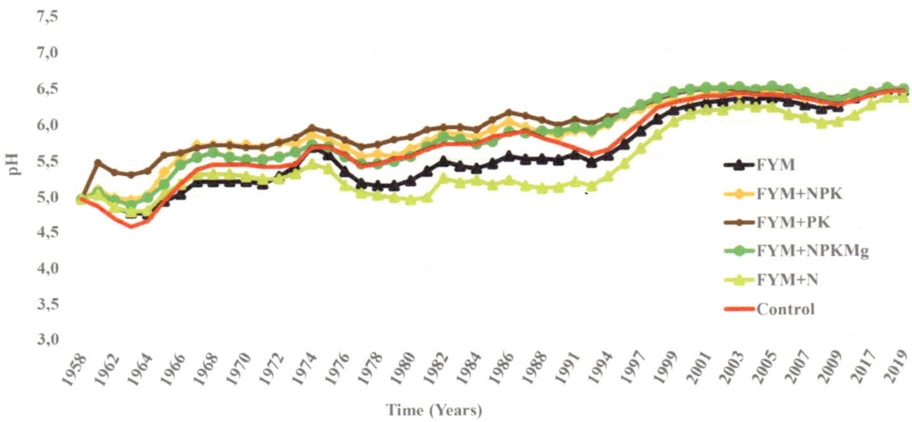

Figure 18.
Effect of integrated nutrition and FYM on soil pH in 0–30 cm depth.

measured in 1958, because crop residues have been removed from the field until 2009. Integrated nutrition increased soil C compared to only FYM application (**Figure 17**). Reduction of crop growth and low root biomass indicated by low yield (**Figure 10**) decreased soil C content at the treatment without any fertilizer. During 1959–1973, the soil organic C was not measured.

Liming of soil every 3 years with 1 tone CaO per ha increased soil pH compared to the initial pH measured in 1958 (**Figure 18**).

4. Discussion

Sustainable crop production can be defined as the agricultural practices that meet human needs for food, is economically viable, while at the same time being environmentally positive [26]. Evaluation of 62 years data of the classical long term trial has shown the importance of nutrient management on all three aspects of sustainability: (1) Improvement of crop yield, sustainable yield index, WUE and soil fertility to produce sufficient food (Social), (2) Profitability of crop production, efficient use of resources, and maintenance of soil fertility to grow the income of farmers (Economic), and (3) Efficient use of nutrients and improving soil fertility to minimize nutrient loss (Economic and Environmental).

4.1 Effect of balanced nutrition on the sustainability of crop production

The data of the trial confirmed that a balanced nutrition increases crop yield, avoids nutrient deficiency, increases nutrient and water use efficiency, protects the environment by minimizing nutrient loss, and improves soil fertility. Chopra et al. (2014) reported similar results [27]. Application of mineral N + P + K + Mg fertilizer as balanced nutrition resulted in the highest yield and income of crop (**Figure 1**). Similar results were reported for potato with N + P + K + Mg + S application [28] and for maize with N + P + K + Mg application [29, 30] as balanced nutrition compared to treatments of unbalanced nutrition. Omitting nutrients reduced crop yield and resulted in an economic loss of between 89 and 812 $\$ha^{-1}$ (**Figure 1**). Dev (1998) and Chander et al. (2012) also reported that omitting nutrients significantly decreased yield and profit of maize, millet, rice, soybean, and wheat [14, 31]. Application of only FYM significantly reduced crop yield and income compared to the balanced nutrition (**Figure 1**). Černý et al. and Dilshad et al. (2010) and Abid et al. (2020) reported a significant reduction of barley, maize, potato, rice, and wheat yields with the application of FYM only compared to N + P + K mineral fertilizer [32–34]. Bhattacharyya et al. (2014) reported a decrease in maize and wheat economic benefit at FYM alone compared to N + P + K fertilizer [35].

Crop varieties improved during the 62 years of the trial and resulted increasing yields. Evaluation of the cereal yield per ha indicates that the average yield was increased by 101% during the latest decade (2010–2020) compared to the average yield measured between 1961 and 1970 for Germany [19]. Fertilizer application is an essential management practices that positively affects yield and income in the long-term. The balanced nutrition of the essential plant nutrients was best nutrient management practice and resulted in the highest yield (**Figures 4** and **5**). Crop yield and income increased during the years 2008 to 2020 were 66% and 1901$\$ha^{-1}$higher than in the earliest years 1958 to1968 (**Figures 4** and **5**). The combined effects of better varieties and fertilizer application improved crop yield and soil fertility during the trial periods (**Figures 2, 3** and **9**). An unbalanced nutrition resulted in yield loss of 6 to 53% or up to 311$\$ha^{-1}$ during the earliest decade, but 8 to 69% (158 to 3075$\$ha^{-1}$) loss during the latest decade. N fertilizer application without PK fertilizer similarly resulted in 86% reduction of maize grain yield during 2001–2006 compared to maize yield at N without PK fertilizer during 1987–1988, because cumulative K releasing power of the soil has depleted 33% in 2003 compared to K releasing power of the soil in 1986 [36]. The high productivity and revenue per area with the balanced nutrition support the social and economic aspects of sustainable crop production.

The long-term application of balanced nutrient positively affects the stability of crop yield [37]. Balanced nutrition (N + P + K + Mg fertilizer) resulted in the highest SYI compared to omitting nutrients (**Figure 6**). Ray et al. (2017) similarly reported the highest SYI with the balanced nutrition (N + P + K + S + Zn fertilizer) compared to the unbalanced nutrition [38]. Application of only FYM decreased SYI by 8% compared to both the N + P + K and N + P + K + Mg treatments (**Figure 6**). Abid et al. (2020), Bhattacharyya et al. (2014), and Malarkodi et al. (2019) reported a reduction of SYI, respectively by 6%, 17% and 5% with only FYM compared to N + P + K mineral fertilizer [34, 35, 39]. The highest SYI was observed at the balanced nutrition and it confirms stable yields as an indicator of sustainability.

Land, solar energy, and water are the major natural resources required to produce crop. Efficient utilization of these resources is necessary for sustainable crop production. Nutrient application is important to increase land and water use efficiency. Reduction of yield because of unbalanced nutrition (**Figure 1**) resulted in inefficient use of land, because more land (1.1 to 1.7 ha) is required to achieve the same yield as on 1 ha of land at the balanced nutrition treatment. The N + P + K + Mg fertilizer resulted in the highest WUE compared to omitting nutrients and FYM application alone (**Figure 6**). Omitting nutrients decreased WUE by 7 to 63%. Chander et al. (2013), Suhas et al. (2013) and Chander et al. (2012) reported similar results [6, 8, 31]. Efficient use of resources is only possible in a balanced application of plant nutrients.

The best practice of nutrient management increases nutrient use efficiency in crop production. The N, P, and K use efficiencies (NUE, PUE, and KUE) of major cereal crops are reported to be between 40 and 65%, 15–25%, and 30–50% respectively at recommended management practices with recommended soil P and K contents [40]. The balanced nutrition (N + P + K + Mg fertilizer) resulted in a high nutrient use efficiency compared to the unbalanced nutrition. The average NUE, PUE, and KUE of crop at the balanced nutrition were, respectively 77%, 49% and 84% (**Figure 7**). Omission of nutrients and application of only FYM decreased nutrient use efficiency by 5–56%, because crop growth and yield were limited by nutrient deficiency. Kumar et al. (2021) reported reduction of nutrient use efficiency by 27–65% for potatoes due to nutrient omission compared to the balanced nutrition [41]. Similar results were reported in Wang et al. (2010) for maize and wheat [42]. Inefficient use of nutrients causes a high cost of production or economic loss and a high risk of environmental pollution.

Physical, chemical, and biological parameters of soil fertility influence the capacity of soil to support crop growth. Nutrient management with its direct impact on nutrient and organic matter contents, pH, and cation exchange capacity of soil supports sustainable crop production. Long-term soil fertility is ensured by balanced nutrition and concurrent application of lime [43]. Omitting P fertilizer decreased soil P content (**Figure 3**). Bhattacharyya et al. (2015) also reported a reduction of soil P content due to P fertilizer omission [44]. K removal without replacement depleted soil K from the medium to the low level (**Figure 2**). Zhao et al. (2014) reported that omitting K fertilizer similarly decreased soil K content at different depths [45]. Balanced nutrition improved soil nutrient content to desirable levels and increased yield with positive impacts on sustainable crop production.

The soil organic matter improves soil water-holding, aeration, nutrient absorption and release, and minimization of leaching and erosion [46]. Application of mineral fertilizer and FYM alone slightly increased soil organic carbon (SOC) compared to the treatment without any fertilizer (**Figure 8**). This was also found by Aula et al. (2016), they reported a significant increase in SOC through the application of NP and NPK fertilizers and FYM compared to without any fertilizer [47]. The only slight increase of organic matter at the application of mineral fertilizer and FYM alone was caused only by root residues (**Figure 1**), because crop residues were removed from the field for more than 50 years. Zhao et al. (2014) reported a significant increase of SOC at mineral NP and NPK fertilizers plus wheat straw compared to NP and NPK fertilizers without straw [45]. The unbalanced nutrition depleted soil organic matter content through a low crop yield. The balanced nutrition improves soil organic matter with positive implications on soil fertility supporting crop growth and yield directly related to sustainable crop production.

4.2 Effect of integrated crop nutrition on the sustainability of crop production

The integrated crop nutrition as the combination of organic and mineral fertilizer contributes: (1) to maintain or enhance soil fertility, (2) to improve nutrient stocks in the soil, and (3) to reduce nutrient loss to the environment by increasing nutrient use efficiency [48]. It improves the availability of nutrient and corrects nutrient imbalances to increase crop yield. Application of FYM plus mineral fertilizers significantly increased crop yield compared to FYM alone (**Figure 10**), which was also reported by Abid et al. (2020) and Mahmood et al. (2017) for maize yield [34, 49] and by Baniuniene and Zekaite (2008) for potato yield [50]. FYM without mineral fertilizer reduced crop yield by 44% and 492$ha^{-1} (**Figure 10**). Bhattacharyya et al. (2014) similarly reported 47% yield and 59% profit reductions for maize and 49% and 52% for wheat at only FYM application compared to FYM plus NPK fertilizer [35]. The integration of FYM with mineral fertilizer increased yield and income, because it improved nutrient availability required to support the healthy growth of crops.

Integrated nutrition was the best nutrient management practice, because it increased crop yield and income to the highest level (**Figures 10–12**) and it improved soil fertility (**Figures 15–17**). Vasuki et al. (2009) similarly reported that the integrated and balanced use of mineral fertilizer plus organic manures have maintained an increase of crop yield at a higher level over the years [36]. Application of only FYM resulted in a loss of income of 1347 $ha^{-1} in the latest years (2008–2020) compared to 237 $ha^{-1} in the earliest years (1958–1968) of the trial, as compared to the treatment of FYM with NK fertilizer during the respective time intervals (**Figure 12**). Hejcman and Kunzova (2010) similarly reported that wheat yield reduction due to application of FYM only was high during the latest decade (1997–2006) and low during the earliest decade (1957–1966) compared to yield at integrating FYM with NPK fertilizer [51]. The synergy between improved varieties and integrated nutrition sustained the increasing yield and income during the long-term, because nutrients have been available in quantity and ratio demanded by high-yielding crop varieties.

The SYI is viewed as a quantitative measurement of sustainability. A high SYI with minimum standard deviation indicates low variability of yield. Integrated nutrition increased SYI compared to the application of FYM alone (**Figure 13**). Integrating FYM with NPK fertilizer similarly increased SYI of maize [34] and sunflower [39] compared to only FYM treatment. Low SYI at only FYM application shows a high variability of yield, while the high SYI at the integrated nutrition indicates sustainable crop production.

Natural resource use efficiency of crop production is increased by improving crop growth. Best nutrient management is therefore needed to achieve efficient utilization of land and water for crop production. Application of only FYM resulted in inefficient land use, because crop yield was 44% lower than at the integrated nutrition (**Figure 10**). Therefore, it requires 1.4 ha of land to achieve the same yield as with FYM plus NP fertilizer on 1 ha, and it decreased WUE by 40% (**Figure 13**). Dubey et al. (2014) similarly reported a 9% reduction of WUE of the crop at only FYM application compared to FYM plus NPK fertilizer [52]. Improvement of land and water use efficiency is an important contribution to sustainable crop production.

Efficient use of nutrients applied as organic plus mineral fertilizers reduces nutrient losses, protects the environment and improves economic return on investment in fertilizer. It was confirmed in the trial data as the highest percentage of crop

NUE, PUE, and KUE were achieved with integrated nutrition. Application of only FYM decreased nutrient use efficiency by 15–24% compared to integrated nutrition (**Figure 14**). Abid et al. (2020) similarly reported a 36% reduction of nutrient use efficiency of maize [34] and Bhattacharyya et al. (2014) reported a 24% and 23% reduction for maize and wheat [35] at only FYM compared to integrating FYM with NPK fertilizer. Application of only FYM was resulted in inadequate and unbalanced availability of nutrients, so that it has been caused a reduction of crop growth and yield, which were ultimately leading to low recovery and inefficient use of nutrients.

Nutrient management improves nutrient availability in the soil and supports soil fertility via its impact on nutrient content, soil organic matter and pH. Integrating FYM with P fertilizer increased soil P_2O_5 content compared to only FYM (**Figure 15**). Malarkodi et al. (2019) and Hejcman and Kunzova (2010) reported similar results [39, 51]. FYM plus K fertilizer-maintained soil K_2O content within the medium range, but only FYM and FYM + NP fertilizer decreased soil K_2O to the 'low' level compared to the initial soil K_2O (**Figure 16**). Application of only FYM similarly depleted soil K_2O compared to FYM plus K fertilizer [39, 49]. Integrated nutrition improved soil nutrient content and increased crop production as an indicator of efficient use of input and resources with positive implications on sustainability.

Some authors claim that the production of cereal crops have stagnated or declined in recent years due to unbalanced and inadequate nutrient application and degradation of the soil organic matter [27]. The decomposition of organic matter releases the nutrients necessary to increase crop yield. Integrated nutrition increased soil organic carbon (SOC) compared to the application of FYM alone (**Figure 17**). A similar result was reported in Malarkodi et al. (2019) and Hejcman and Kunzova (2010) [39, 51]. An increase in SOC indicates organic matter improvement that makes soil condition favorable to increase yield and to sequestrate carbon in crop residues.

The soil pH regulates solubility and availability of nutrients. It increased rapidly during 1958–1998 at integrated nutrition compared to FYM alone by CaO (lime) application (**Figure 18**). Abid et al. (2020) similarly reported that supplementing FYM with NPK fertilizer significantly increased soil pH compared to only FYM [34]. Since 1998, the soil pH was maintained at a desirable level with a slight difference between treatments due to the accumulated effect of lime.

5. Conclusion

Analysis of 62 years of data of the long-term trial confirmed that application of mineral fertilizer N + P + K + Mg as the balanced nutrition and supplementing FYM with mineral fertilizer as the integrated nutrition supports the social, economic, and environmental aspects of sustainable crop production. Any unbalanced nutrition caused by omitting nutrients or applying average quantities of FYM alone resulted in a reduction of crop yield and revenue. It contributed to inefficient use of nutrients and resources, an unstable yield increase, and a depletion of soil fertility with negative implications on sustainability.

Violation of the Law of the Minimum by omitting nutrients decreased crop yield, revenue, SYI, WUE, NUE, PUE, and KUE, respectively by 7–65%, 89–812 $\$ha^{-1}$, 1–22%, 7–63%, 5–23%, 6–49%, and 7–56% compared to the balanced nutrition, because essential functions of the missing nutrients cannot be fulfilled by any other nutrient. Application of FYM alone as organic fertilizer at the local rates in the long-term trial decreased crop yield, revenue, SYI, WUE, NUE, PUE, and KUE, respectively by 44%, 492 $\$ha^{-1}$, 10%, 40%, 33%, 23%, and 24% compared to the integrated nutrition, because nutritional needs of crop were not fully satisfied due to unpredictable availability and the unbalanced ratio of nutrients in the FYM.

Therefore, both the balanced and integrated principles of crop nutrition are the best management strategies to support the positive impacts of technological progress in crop production without depleting the soil fertility. They are important to sustain crop production for future generations while the environment is protected.

Acknowledgements

Many thanks to the field trials manager, Klemens Brüggemann, and his team for the collection and organization of plant and soil samples of the long-term trial. Thanks to the staff at Hanninghof laboratory for the analysis of the samples. Without their support, it would not have been possible to come up with the evaluation of the data to prepare the results for publication.

Author details

Melkamu Jate* and Joachim Lammel
Research Centre Hanninghof, Yara International ASA, Duelmen, Germany

*Address all correspondence to: melkamu.jate@yara.com

References

[1] Steiner RA, Herdt RW. A global directory of long-term agronomic experiments: Non-European experiments. Vol. l. New York, NY: Rockefeller Foundation; 1993

[2] McRae KB, Ryan DAJ. Design and planning of long-term experiments. Canadian Journal of Plant Science. 1996;**76**:595-602

[3] FAO, Organization of Economic Co-operation and Development (OECD). Food security and nutrition: Challenges for agriculture and the hidden potential of soil [Internet]. A report to the G20 agriculture deputies. 2018. Available from: http://www.fao.org/3/CA0917EN/ca0917en.pdf [Accessed: November 10, 2021]

[4] Roy RN, Finck A, Blair GJ, Tandon HLS. Plant Nutrition for Food Security, a Guide for Integrated Nutrient Management. Rome: FAO Fertilizer and Plant Nutrition Bulletin 16; 2006. p. 17

[5] Peltonen-Sainio P, Jauhiainen L, Laurila IP. Cereal yield trends in northern European conditions: Changes in yield potential and its realization. Field Crops Research. 2008;**4960**:1-6. DOI: 10.1016/j.fcr.2008.07.007

[6] Chander G, Suhas PW, Kanwar L, Pal CK, Mathur TP. Integrated plant genetic and balanced nutrient management enhances crop and water productivity of rainfed production systems in Rajasthan, India. Communications in Soil Science and Plant Analysis. 2013;**44**: 3456-3464

[7] Vyn T. Boosting global corn yields depends on improving nutrient balance [Internet]. 2014. Available from: https://extension.purdue.edu/article/6584 [Accessed: November 10, 2021]

[8] Suhas PW, Chander G, Sahrawat KL, Dixit S, Venkateswarlu B. Improved crop productivity and rural livelihoods through balanced nutrition in the rainfed semiarid tropics. International Crops Research Institute for the Semi-Arid Tropics, Patancheru, Andhra Pradesh and Indian Council of Agricultural Research, New Delhi, India. Resilient Dryland Systems Report number 58. 2013. p. 2

[9] Alley M. M, Vanlauwe B. The role of fertilizers in integrated plant nutrient management [Internet]. IFA and Tropical Soil Biology and Fertility Institute of the International Centre for Tropical Agriculture. 2009. Available from: www.fertilizer.org/ifacontent/.../2/file/2009_ifa_role_plant_nutrients.pdf [Accessed: November 10, 2021]

[10] Körschens M. The importance of long-term experiments for soil science and environmental research – A review. Plant Soil Environment. 2006;**52**(Special issue):1-8

[11] Goulding K. Long-term research in the UK: Lessons learned from the Rothamsted classical experiments. In: Report from a Conference on Success Stories of Agricultural Long-Term Experiments. 28-29 May 2007. Uppsala, Sweden: The Royal Swedish Academy of Agriculture and Forestry; 2007. pp. 7-12

[12] Gruhn P, Goletti F, Yudelman M. Integrated Nutrient Management, Soil Fertility, and Sustainable Agriculture: Current Issues and Future Challenges. Vol. 32. Washington DC, USA: International Food Policy Research Institute; 2000. pp. 1-38

[13] International Potash Institute (IPI). Balanced fertilization for sustaining crop

productivity. In: Benbi DK, Brar MS, Bansal SK, editors. Proceedings of the International Symposium; 22-25 November 2006; India. Ludhiana. Horgen, Switzerland: International Potash Institute; 2006. pp. 8-10

[14] Dev G. Balanced fertilizer use increases crop yield and profit in India. Better Crops International. 1998;**12**(2):25-28

[15] Fertilizer Europe (FE). Balanced plant nutrition [Internet]. 2019. Available from: https://www.fertilizerseurope.com/fertilizers-in-europe/balanced-plant-nutrition/ [Accessed: 2021 November 10]

[16] Jones DL, Healey JR. Organic amendments for remediation: Putting waste to good use. Elements. 2010;**6**:369-374

[17] Yara and KTBL. Faustzahlen für die Landwirtschaft. Darmstadt, Germany: KTBL-Schriftenvertrieb im Landwirtschaftsverlag; 2005. p. 263

[18] Kurt GB Lexikon des Agrarraums [Internet]. 2021. Available from: https://www.agrarraum.info/lexikon-g.html#getreide [Accessed: November 10, 2021]

[19] FAO: FAOSTAT. Production crops: data of crop yield and price for Germany [Internet]. 2021. Available from: http://www.fao.org/faostat/en/#data/QCL [Accessed: October 7, 2021]

[20] Yara. Historical data of mineral fertilizer price. "Unpublished data". 2020

[21] Singh PR, Rao SK, Bhaskarrao DUM, Ready MN. Sustainability Index under Different Management; Annual Report. Hydarabad, India: CRIDA; 1990

[22] Qaswar M, Huang J, Ahmed W, Li D, Liu S, Ali S, et al. Long-term green manure rotations improve soil biochemical properties, yield sustainability and nutrient balances in acidic paddy soil under a rice-based cropping system. Agronomy. 2019;**9**:780. DOI: 10.3390/agronomy9120780

[23] Sharmaa A, Sankar GRM, Arorac S, Guptaa V, Singha B, Kumar J, et al. Analyzing rainfall effects for sustainable rainfed maize productivity in foothills of Northwest Himalayas. Field Crops Research. 2013;**145**:96-105

[24] Dobermann A. Nutrient use efficiency measurement and management. In: Papers presented at the IFA international workshop on fertilizer best management practices; 7-9 March 2007; Belgium. Brussels. Paris, France: International Fertilizer Industry Association; 2007. pp. 7-8

[25] Landwirtschafts-kammer Nordrhein-Westfalen (LKNRW): Düngung mit Phosphat, Kali, Magnesium [Internet]. 2015. Available from: https://www.landwirtschaftskammer.de/landwirtschaft/ackerbau/pdf/phosphat-kalium-magnesium-pdf.pdf [Accessed: November 10, 2021]

[26] Allen GG, Begam R, Shrawat A. Sustainable crop production. In: Hudson RJ, editor. Physiology, Biochemistry and Molecular Biology in Animal and Plant Productivity. UK: Encyclopedia of Life Support Systems (EOLSS) and UNESCO, Eolss publishers Co. Ltd; 2010. pp. 294-313

[27] Chopra R, Singh A, Katara P. Nutrient management: Key for sustainable crop production. Popular Kheti. 2014;**2**(2):61-64

[28] Zengin M, Gökmen F, Gezgin S, Cakmak I. Effects of different fertilizers with potassium and magnesium on the

yield and quality of potato. Asian Journal of Chemistry. 2008;**20**(1):663-676

[29] El-Dissoky RA, Al-Kamar FA, Derar RM. Impact of magnesium fertilization on yield and nutrients uptake by maize grown on two different soils. Egyptian Journal of Soil Science. 2017;**57**(4):455-466

[30] Ely EO, Sofyan ET, Sara DS. The effect of NPK+Mg fertilizer application on potassium availability, potassium uptake, and yield of sweet corn (Zea mays Saccharata Sturt) in Inceptisols. International Journal of Energy and Environmental Science. 2020;**5**(3):47-50

[31] Chander G, Suhas PW, Sahrawat KL, Jangawad LS. Balanced plant nutrition enhances rainfed crop yields and water productivity in Jharkhand and Madhya Pradesh states of India. Journal of Tropical Agriculture. 2012;**50**(1-2):24-29

[32] Černý J, Balík J, Kulhánek M, Čásová K, Nedvěd V. Mineral and organic fertilization efficiency in long-term stationary experiments. Plant Soil Environment. 2010;**56**(1):28-36

[33] Dilshad MD, Lone MI, Jilani G, Malik MA, Yousaf M. Integrated plant nutrient management on maize under rainfed condition. Pakistan Journal of Nutrition. 2010;**9**(9):896-901

[34] Abid M, Batool T, Siddique G, Ali S, Binyamin R, Shahid MJ, et al. Integrated nutrient management enhances soil quality and crop productivity in maize-based cropping system. Sustainability. 2020;**12**(23):10214. DOI: 10.3390/su122310214

[35] Bhattacharyya R, Pandey AK, Gopinath KA, Mina BL, Bisht JK, Bhatt JC. Fertilization and crop residue addition impacts on yield sustainability under a rainfed maize–wheat system in the Himalayas. Proceeding of the National Academy of Sciences, India, Section B: Biological Sciences. 2014;**81**(6). DOI: 10.1007/s40011-014-0394-8

[36] Vasuki N, Yogananda SB, Preethu DC, Sudhir K. Impact of Long Term Fertilizer Application on Soil Quality, Crop Productivity, and Sustainability–Two Decades Experience. Bangalore; New Delhi, India: University of Agricultural Sciences; ICAR; 2009. p. 6

[37] Ahrends HE, Siebert S, Rezaei EE, Seidel SJ, Hüging H, Ewert F, et al. Nutrient supply affects the yield stability of major European crops—A 50 year study. Environmental Research Letters. 2021;**16**:014003. DOI: 10.1088/1748-9326/abc849

[38] Ray M, Haldar P, Saha S, Chatterjee S, Adhikary S, Mukhopadhyay SK. Effect of balanced nutrition on productivity, economics and soil fertility of rice (Oryza sativa L.) – greengram [Vigna radiata (L.) Wilczek] cropping system under coastal West Bengal. Journal of Crop and Weed. 2017;**13**(1):89-92

[39] Malarkodi M, Elayarajan M, Arulmozhiselvan K, Gokila B. Long-term impact of fertilizers and manures on crop productivity and soil fertility in an alfisol. The Pharma Innovation Journal. 2019;**8**(7):252-256

[40] Fixen P, Brentrup F, Bruulsema TW, Garcia F, Norton R, Zingore S. Nutrient/fertilizer use efficiency: Measurement, current situation and trends. In: Pay D, Patrick H, Hillel M, Robert M, Dennis W, editors. Managing Water and Fertilizer for Sustainable Agricultural Intensification. Paris, France: IFA; 2015, 2015. pp. 8-39

[41] Kumar P, Dwivedi DK, Bharati V, Tigga A, Singh H, Dwivedi A. Response

of NPK on growth and yield of potato (Solanum tuberosum L.) under calcareous soils of Bihar. International Journal of Current Microbiology App Science. 2021;**10**(02):1956-1961

[42] Wang Y, Wang E, Wang D, Huang S, Ma Y, Smith CJ, et al. Crop productivity and nutrient use efficiency as affected by long-term fertilization in North China Plain. Nutrient Cycle Agroecosystem. 2010;**86**:105-119

[43] Thompson LM, Troeh FR. Soils and soil fertility. 3rd ed. New York: McGraw-Hill; 1973

[44] Bhattacharyya P, Nayak AK, Shahid M, Tripathi R, Mohanty S, Kumar A, et al. Effects of 42-year long-term fertilizer management on soil phosphorus availability, fractionation, adsorption-desorption isotherm and plant uptake in flooded tropical rice. The Crop Journal. 2015;**3**(5):387-395

[45] Zhao S, He P, Qiu S, Jia L, Liu M, Jin J, et al. Long-term effects of potassium fertilization and straw return on soil potassium levels and crop yields in north-central China. Field Crops Research. 2014;**169**:116-122

[46] Sekhon GS, Meelu OP. Organic matter management in relation to crop production in stressed rain fed systems. In: Virmani SM, Katyal JC, Eswaran H, Abrol IP, editors. Stressed Ecosystems and Sustainable Agriculture. New Delhi: Oxford University Press and IBH Publishing; 1994

[47] Aula L, Macnack N, Omara P, Mullock J, Raun W. Effect of fertilizer nitrogen (N) on soil organic carbon, total N and soil pH in long-term continuous winter wheat (*Triticum Aestivum* L.). Communications in Soil Science and Plant Analysis. 2016;**47**(7):863-874

[48] FAO. Guide to efficient plant nutrition management [Internet]. Land and water development division of FAO of the UN. 1998. Available from: www. ftp://ftp.fao.org/agl/agl/docs/gepnm.pdf [Accessed: November 10, 2021]

[49] Mahmood F, Khan I, Ashraf U, Shahzad T, Hussain S, Shahid M, et al. Effects of organic and inorganic manures on maize and their residual impact on soil physico-chemical properties. Journal of Soil Science and Plant Nutrition. 2017;**17**(1):22-32

[50] Baniuniene A, Zekaite V. The effect of mineral and organic fertilizers on potato tuber yield and quality. Agronomijas Vestis (Latvian Journal of Agronomy). 2008;**11**:2002-2006

[51] Hejcman M, Kunzova E. Sustainability of winter wheat production on sandy-loamy Cambisol in the Czech Republic: Results from a long-term fertilizer and crop rotation. Field Crops Research. 2010;**115**:191-199

[52] Dubey R, Sharma RS, Dubey DP. Effect of organic, inorganic and integrated nutrient management on crop productivity, water productivity and soil properties under various rice-based cropping systems in Madhya Pradesh. India. International Journal of Current Microbiology App Science. 2014;**3**(2):381-389

Chapter 3

Vermicomposting: A Step towards Sustainability

Priyanka Saha, Anamika Barman and Anurag Bera

Abstract

Agricultural production depends on so many things. Proper nutrient management is one of them. It becomes a trend to apply excess amount of fertilizer for enhancing productivity without considering its effect on soil health. Vermicomposting is a process of scientifically decomposing agricultural, municipality, and industrial wastes into nutrient enriched compost by earthworms. Vermicompost not only balance underground soil environment and makes is a suitable habitat for soil micro biota but also improves above ground environment. Microbes are the fundamental element of ecosystem. Use of vermicompost increases growth and proliferation of microbes that amplify environment's betterment. Vermicomposting is also affordable for resource poor small and marginal farmers. Therefore, vermicompost use is more economical than synthetic organic fertilizer. So, economic viability, environmental stability, and enhancing livelihood quality are the major causes for its worldwide adoption in food production.

Keywords: vermicompost, soil fertility, sustainability, earthworm, soil health

1. Introduction

Increasing population and food demand has forced the farming community to apply excess amount of chemical fertilizer that leads to degradation of soil health and causing environmental pollution. Factor productivity of the soil is also decreasing due to injudicious fertilization. The technology advancement and industrialization has created many challenges associated with sustainability. Sustainability is a concept of utilizing the natural resources without compromising the ability of future generation to meet their own needs. Rapid urbanization and industrial growth is worrisome with respect to huge amount of waste generation. Unscientific management of these wastes causing social, economic, and environmental problems. After consuming so much chemicals during the green revolution era, the soil eventually became unproductive due to a lack of sufficient organic matter amendments. Vermicomposting is one of the many potential approaches that have gained significant attention over decades. It is an eco-friendly concept of waste management where decomposition process is aided by microorganisms [1–3]. Earthworms are the biological engineers since the beginning of humankind. The technique of culturing earthworm for managing wastes and preparing compost is known as vermicomposting. Vermicomposting is defined as a bio-oxidative process where earthworms and decomposer microorganisms (bacteria, fungi, and actinomycetes) act synergistically to manage organic waste in a scientific way that also aids in improvement of soil physical, chemical, and biological properties [4]. A wide range of raw materials (**Figure 1**) such as agricultural waste [5], animal waste [6], and municipality [7] waste are decomposed by earthworms and microorganisms for preparing vermicompost. This bio-technique increases mineralization of waste material led to enhancement in bioavailability of essential plant nutrients. Vermicompost not only supplies plant nutrients and growth promoting hormones but also improves soil physical property through soil aggregation [8]. Hence it is used as a component of organic farming. Vermicompost has also been proven to be a miraculous plant growth stimulator [9]. Vermicast, the end product is also rich in hormones and enzymes which make the soil environment favorable for soil biota. Residue burning is a common issue nowadays that causing severe environmental hazards. This issue can also be overcome by adopting vermicomposting technique.

In spite of having so many benefits, use of vermicompost is still not accepted widely due to lack of awareness and technology barriers. There is a need for proper extension to explore the potentialities of vermicompost. So, this study was conducted with the objectives for getting a precise idea about general properties, preparation methods, benefits and its limitations, and most importantly understanding the significance of vermicompost in crop production.

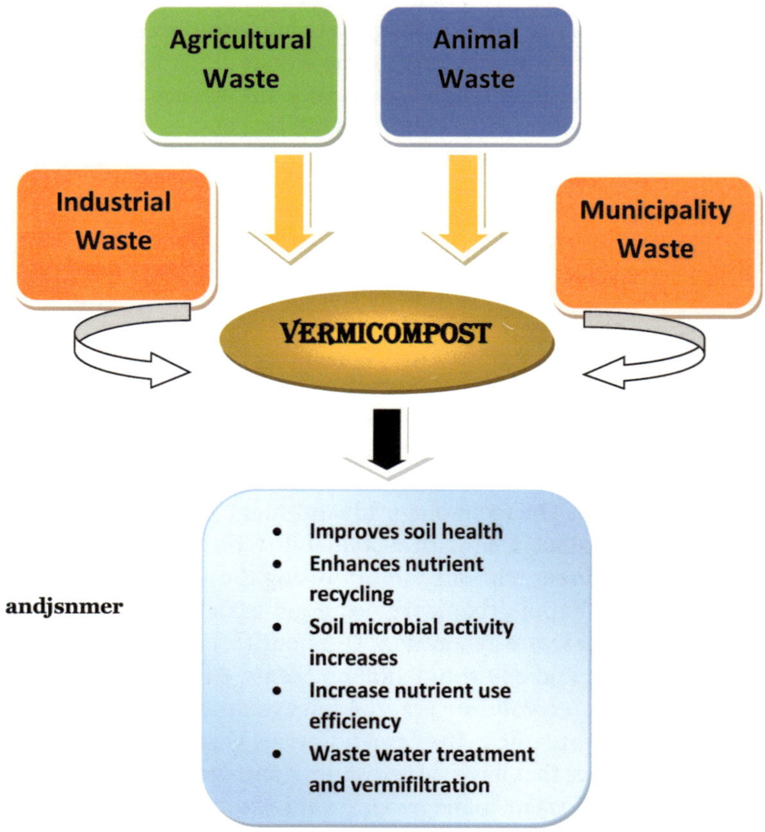

Figure 1.
Vermicompost and its role in agriculture.

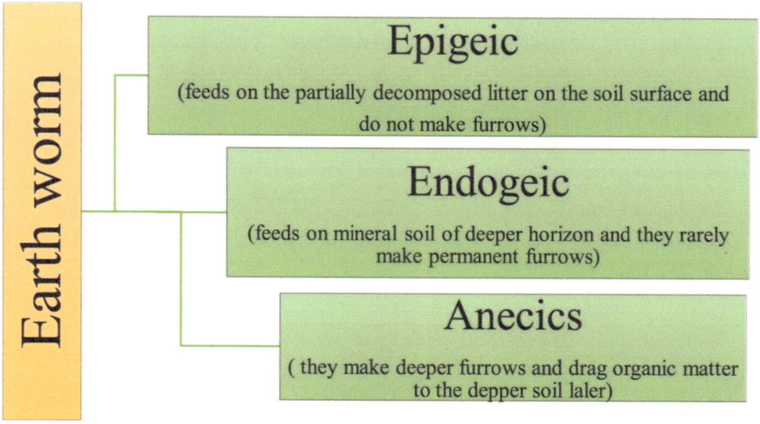

Figure 2.
Earth warm classification.

2. Earth worm and its kind

Among the soil biota, earth worm is one of the major kinds and a key component of tropical and subtropical ecosystems [10, 11]. It helps is soil aggregation, nutrient recycling, litter decomposition, etc. Earthworm improves the soil environment by producing cast, pellets, and galleries. Mucus secretion from the gut of earth worm enhances microbial activity. Around 3000 species of earthworms documented so far [12]. The earthworms are of three types that have been described in **Figure 2**.

The most common earthworms [13] have successfully used in India for vermicompost preparation are:

Indices	Chemical fertilizers	Vermicompost
Synthesis process	They are synthesized and manufactured in factories.	They are the product of natural decomposition of organic matter with the help of earth worms.
Macronutrients	Major chemical fertilizer contains only one macronutrient (either nitrogen or phosphorus or potassium).	Vermicompost contains almost all the primary minerals along with some quantities of secondary minerals (Ca, Mg, and S) [14].
Micronutrients	Not present.	Significant amount of micronutrients: Zn, B, Mn, Fe, Cu, etc. also present [15].
Soil structure	Over use of chemical fertilizer degrades soil structure.	It improves soil aggregation, water holding capacity, soil aeration, etc.
Biological activity of soil	It reduces biological activity of soil.	It improves activity of soil microbes thus enhances soil fertility [16].
Environmental impact	Excessive use of chemical fertilizers causes environmental pollution.	Vermicompost is an eco-friendly approach [17].
Saving cost of cultivation	Use of chemical fertilizers increases the cost of cultivation.	The farmer/consumer can expect approximately \$110–\$350 in additional income from applying one ton of vermicompost due to offset costs of traditional fertilizer and pesticides [18].

Table 1.
Difference between chemical fertilizer and vermicompost.

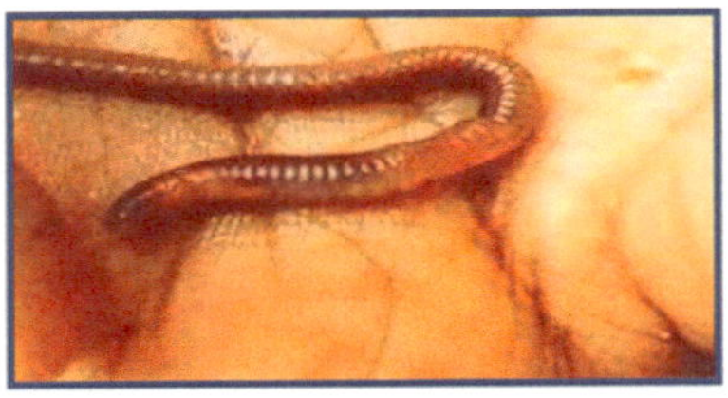

Figure 3.
*African earthworm (*Eudrilus eugeniae*).*

Figure 4.
*Tiger worm (*Eisenia fetida*).*

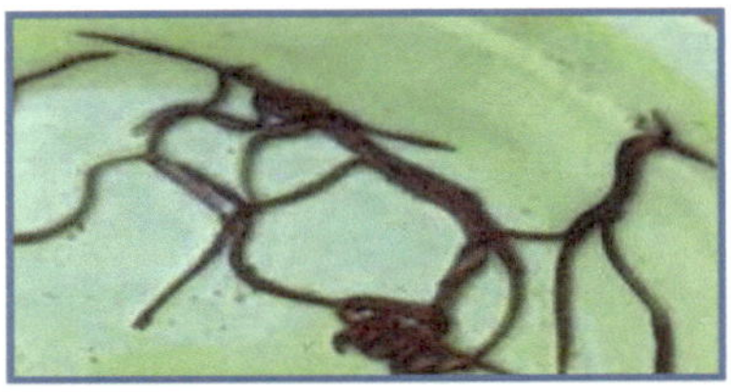

Figure 5.
*Asian worms (*Perionyx excavatus*).*

- *Perionyx excavates* (a native species)
- *Eisenia fetida* (exotic species that have colonized many ecosystems)
- *Eudrilus eugeniae* (exotic species largely confines to experimental setup)

Apart from being ecological engineer, earth worm is a rich source of protein thus it can be used as high quality feed to farm animals. Das et al. [13] reported that earthworm cast increases mushroom production. The brief difference between chemical fertilizer and vermicompost is given in **Table 1**.

Most commonly used earthworm species are: African earthworm (*E. eugeniae*), that is, **Figure** 3, Tiger worm (*E. fetida*), that is, **Figure** 4, and Asian worms (*Perionyx excavatus*), that is, **Figure** 5.

3. General properties of vermicompost

In terms of sustainable crop production, the acceptability of vermicompost has been rising rapidly as soon as the human realizes the significance of organic inputs in crop field. The excreta of earthworms, which is considered as the main product, that is, vermicompost has several characteristics. These are:

3.1 Physical properties

- A good vermicompost is always non-toxic, well-decomposed, ecologically compatible, and environment friendly.
- Any type of green waste viz. municipal waste, agricultural waste, sewage sludge, industrial waste, and human feces can be used for the conversion by earthworm.
- When turning of soil is occurred in proper manner, it is symptomatic to aerobic decomposition which will produce normal odor after preparation. If there is improper aeration, foul odor can be formed.
- The final outcome of vermicomposting would be comprising of fine particulate structure, granular form.
- Vermicompost plays the role of a "soil conditioner" by improving the soil porosity, drainage, and water holding capacity [19].

3.2 Chemical properties

- Vermicompost is rich in almost all essential macro and micro plant nutrients. Several experiment states that average nutrient content of vermicompost is greater than other conventional compost, produced from other procedures.
- Among all the secondary nutrients, calcium content in vermicompost is higher than other compost.
- In contrast with other conventional compost, vermicompost contains worm mucus which facilitates in preventing washing away of nutrients present there [20].
- Due to vermi-conversion, heavy metal present in feeding material is found to be reduced in earthworm cast owing to its accumulation in worm tissue. According to the feed used, the rate of removal of heavy metal depends in vermicomposting techniques. This property makes vermicompost lesser contaminant than any other compost. Thus, it becomes more environmentally sustainable [21].
- There are certain differences found in chemical properties between simple farm yard compost and vermicompost. Vermicompost ranges higher in macro and micro-nutrients as well as soil organic carbon status that can be observed from the **Table 2** [22].

3.3 Biological properties

- The by-product of earth casting is an inhabitant of several microorganism, viz. bacteria, fungi, and actinomycetes. These micro-organisms release several enzyme and phytohormones which helps in improving plant growth. Thus, vermicompost facilitates both microbial and enzymatic activity [22, 23].
- The microbial population of nitrogen fixer bacteria and other symbiotic associative bacteria are supposed to be in a good range of numbers in the excreta of earthworm.
- In addition, earthworm casts harbor a large number of vesicular-arbuscular mycorrhiza (VAM) propagules. These propagules survive up to 11 months on the

Properties	Compost	Vermicompost
pH	7.16	7.72
EC (dSm^{-1})	3.65	6.88
OC	20.5	17.3
Total N (%)	2.42	3.5
Total P (%)	0.88	0.71
Total K ($mg.kg^{-1}$)	653.5	950.5
Total Ca (%)	2.9	3.5
Total Mg (%)	1.5	2.8
Total Fe ($mg.kg^{-1}$)	4467	6045
Total Zn ($mg.kg^{-1}$)	115.5	189.5
Total Cu ($mg.kg^{-1}$)	59	38
Total Mn ($mg.kg^{-1}$)	221.45	344.15
C:N	8.47	5.51

Table 2.
Chemical properties of compost and vermicompost.

Properties	Impact	References
Soil physical properties	Soil aggregation, soil structure, and water holding capacity, infiltration rate improves after vermicompost application.	Edwards and Burrow [19]
Soil chemical properties	Vermicompost also offers a greater chance for reducing salinity, alkalinity, and reduction of heavy metal contamination.	Nancarrow et al. [20]
Soil microbial properties	Microbial biomass is also increases with the use of vermicompost.	Blouin et al. [12]

Table 3.
Effect of vermicompost on different soil properties.

cast, and helps in increasing microbial activity to produce nitrogen and phosphorus in readily available form to the plant (**Table 3**) [24].

4. Preparation methods

Earthworms are often termed as "Bio-engineers" because of their unique ability to convert organic wastes into dark brown nutrient rich compost materials. We use these worms along with some easy-available inputs to produce the vermicompost. In South-Asian countries like India, we often see market price of the vermicompost is very low, which is attributed to the low-cost inputs of this compost. This vermicompost can be prepared in various techniques, among all those two most common methods are: bed and pit methods.

Bed method is easy to prepare and maintain throughout the process as here composting is done on pucca or kachcha floor by making the bed with organic materials like hay, straw, corn silage, etc.

DOI: http://dx.doi.org/10.5772/intechopen.102641

Figure 6.
Bed method.

Figure 7.
Pit method.

Figure 8.
Spraying of water in bed.

Figure 9.
Adult worms in compost.

Pit method is comparatively strenuous process where composting is done on cemented pits of approx. The unit is covered with grass or any other organic mixtures (**Figures 6–10**).

Figure 10.
Fully prepared vermicompost.

4.1 Step by step of preparation methods

- Selection of suitable earthworm species. (Surface dwelling earthworms are only preferred)
- Choosing the site of vermicomposting unit in a cool and shady site.
- Organic waste with cowdung and chopped dried leafy materials are mixed in the proportion of 3: 1 ratio. A compost heap is structured and allowed to decompose for 15 to 20 days.
- Vermiculture bed of 3 cm is to be prepared by placing after saw dust or husk or coir waste or sugarcane trash in the bottom of tub / container.
- A layer of fine sand (3 cm) should be spread over the culture bed followed by a layer of garden soil (3 cm). All layers must be moistened with water.
- Partial decomposed materials obtained from first step is to be laded over the bed. Water is to be sprayed over it to make the moisture availability up to 50%.
- Adult earthworms are released @10/kg in the upper layer of the bed.
- Beds should be kept moist by sprinkling of water (daily) and by covering with gunny bags/polythene. Earthworms should remain undisturbed for it's multiplication.
- Bed should be turned once after 30 days for maintaining aeration and for proper aerobic decomposition.
- The fully prepared vermicompost looks dark brown colored granules, appeared like a handful of dry CTC tea.
- The vermicompost is fully prepared within 45-50 days. When raw material is completely decomposed it appears black and granular. Then it is sieved further and should be separated from any contamination before use in crop field

5. Beneficial effects of vermicompost

5.1 Effect of vermicompost on the soil physiochemical properties

Addition of vermicompost improves soil physico-chemical properties viz. soil structure, soil water holding capacity, penetration resistance, bulk density, soil organic carbon, aggregation, nutrient content, etc. According to the findings of various long term research addition of vermicompost reduces the bulk density of the soil and increases the water holding capacity of soil [25]. Aksakal et al. [26] found that when vermicompost was added in the soil, the mean bulk density, and mean total porosity were the least. Air permeability rose and penetration resistance reduced dramatically as wet aggregate stability improved and bulk density reduced. Increased microbial population and activity led in the development of aggregates and increased soil porosity, resulting in decreased particle and bulk densities. Physicochemical characteristics such as pH, electrical conductivity (EC), porosity, moisture content, water holding capacity, and chemical properties like nitrogen, phosphorous, potassium, calcium, and magnesium were all found to be significantly improved in vermicompost treated soil, while the corresponding physicochemical values in control soil were minimal in rice crop [27]. Vermicompost has indeed been found to have significant concentration of total and bioavailable nitrogen, phosphorus, potassium (NPK), and micronutrients, as well as microbial and enzyme activity and growth regulators [28]. Polysaccharides appeared to be abundant in vermicompost [29]. Polysaccharide worked as a cementing ingredient in the soil, causing aggregate stability, which helped to establish and maintain the soil structure for improved aeration, water retention, drainage, and aerobic conditions. The preservation of soil structure is essential for root elongation and nutrient uptake. The inclusion of mucus secretion and microorganisms from the earthworm's gut improves the soil's aggregate stability. The absorbent organic matter in vermicomposts increases the soil's water retention capacity by holding only the quantity of water required by the plant roots [30]. Vermicomposts have been found to have a higher base exchange capacity and a higher oxidation potential rise [31]. The C/N ratio of vermicompost is usually lower, indicating that it is more suited for use as a soil amendment. By altering the physiochemical parameters of the soil, vermicompost was able to limit the loss of nutrients through leaching [32]. Humic acid and biologically active compounds like plant growth regulators are abundant in vermicompost [33]. Humic acid has been proven to improve nutrient accretion in situations where nutrients are scarce or when additional nutrients are provided. Humic acids may have a hormone-like effect on plant growth and productivity as a result of their involvement in cell respiration, photosynthesis, oxidative phosphorylation, biogenesis, and a variety of other enzymatic functions.

5.2 Effect of vermicompost on the soil biological properties

Biological properties of soil can be enhanced through application of vermicompost. Recent studies founded that soil biological characteristics viz. soil organic carbon as well as soil microbial biomass, enzymatic activity, population of different beneficial microorganism, hormones, etc. significantly enhanced with application of vermicompost [34]. The activity of the dehydrogenase enzyme, which is commonly employed to quantify the respiratory activity of microbial communities, was shown to be higher in vermicompost than in commercial medium [35]. Application of vermicompost

Crop	Treatments	Physiochemical effects				References
		pH	EC (dSm^{-1})	BD ($g\ cm^{-3}$)	Porosity (%)	
Rice	Control	7.4 ± 2.01	2.0 ± 1.0	—	39 ± 2.0	Tharmaraj et al. [27]
	Vermicompost	7.1 ± 0.01	1.01 ± 1.0	—	41 ± 1.0	
	Vermi-wash	7.2 ± 1.02	2.0 ± 1.1	—	40 ± 1.1	
	Vermicompost+ Vermi-wash	7.0 + 0.03	0.02 ± 0.01	—	44 ± 1.0	
Wheat	Soil sample	8.56	25.82	1.52	25.38	Mahmoud et al. [36]
	Vermicompost @5 g kg^{-1} soil	7.6	4.65	1.42	26.85	

Table 4.
Effect of vermicompost on physiochemical properties of soil on different crops.

Parameters	Compost ($g\ m^{-2}$)			
	Vermicompost		Conventional compost	
	100	150	100	150
Nitrogen (%)	0.61	0.72	0.54	0.62
Phosphorus (%)	0.0057	0.0077	0.0039	0.0047
Potassium (%)	11.11	11.17	10.41	10.48
Calcium (%)	1.443	1.683	0.561	0.641
Source: Islam et al. [37].				

Table 5.
Comparison between the effect of vermicompost and conventional compost on different nutrient content of the Amaranthus viridis production.

improved the nitrogen status of soil by introducing the beneficial microorganism in the rhizosphere of the plant which ultimately enhances the nitrogenase activity in soil, which is the enzyme responsible for nitrogen fixation (**Tables 4** and 5).

5.3 Effect of vermicompost on the soil fertility

Vermicompost has a great importance to increase the soil fertility level. In recent years organic amendments are getting more importance for nutrient management and sustainable crop production since the long-term use of inorganic fertilizer lacking organic additives has the ability to ruin soil qualities [34]. Long-term treatment of balanced inorganic fertilizers led to reduced soil bulk density, improved total porosity, and higher water-holding capacity. Inorganic fertilizers also promoted soil aggregation in deeper soil layers and raised maize and wheat grain and straw yields [38]. In their research, using farmyard manure (organic fertilizer) instead of inorganic fertilizer improved soil qualities in a similar way. Furthermore, compost provides substantially higher boosts in soil organic carbon as well as some plant nutrients when compared to mineral fertilizers [39, 40]. Thus, using vermicompost improves overall soil fertility by improving numerous soil physical, chemical, and biological qualities.

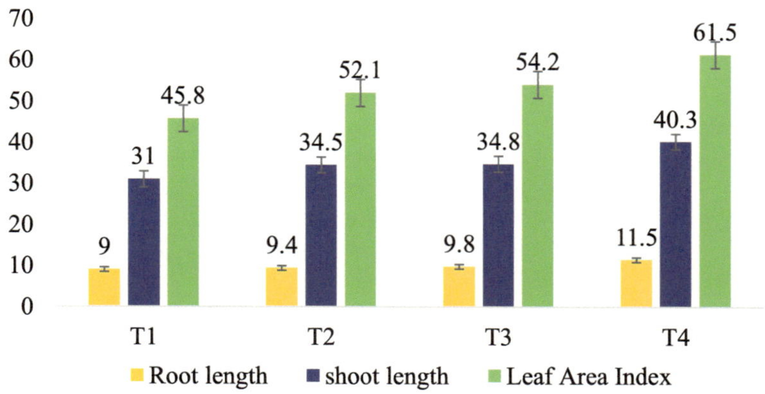

Figure 11.
Effect of vermicompost on growth parameters of Phaseolus vulgaris (20 DAS). Source: Ref. [34]. T1: Control (without application of inorganic NPK or vermicompost), T2: 100% recommended dose of NPK (20:80:40 kg ha^{-1}), T3: 100% recommended dose of vermicompost (5 t ha^{-1}), and T4: 50% vermicompost supplemented with 50% NPK (W/W).

5.4 Effect of vermicompost on plant growth and development

Vermicompost promotes the growth and development of a variety of plant species, especially various horticulture crops, that is, sweet corn, tomato, strawberry [41], cereals crop rice [27], wheat, sorghum [32], fruit crops papaya [42], and pineapple [43]. Several growth and yield metrics viz. stem diameter, plant height, marketable yield per plant, mean leaf number, and total plant biomass of tomato plant were recorded significantly higher with the application of vermicompost (**Figure 11**).

The increase in growth and development of plant is due to the improving action of vermicompost application on soil physical, chemical, and biological properties which ultimately improves the overall soil fertility, which enhances the plant growth and development. Vermicompost has been demonstrated to improve plant dry weight [44] and uptake of plant N [45] serve as a naturally available, slow released sources of plant nutrients.

5.5 Effect of vermicompost on plant diseases

Various studies had showed that vermicompost is useful for remedies of different plant diseases. Many plant diseases caused by soil-borne, foliar plant pathogens, and pests have been suppressed by vermicompost products, which have been proven to be effective as organic fertilizers and biological control agents. In conventional agriculture, excessive and repeated use of chemical pesticides resulted in "biological resistance" in crop diseases and pests. As a result, significantly higher doses are now needed to inhibit them for the growth of high-yielding crops that are more sensitive to pests and diseases [46]. A study was conducted to compared the inhibition performance of two different methods, in which two nonconventional chemicals $ZnSO_4$ and oxalic acid, as well as the bio-control agent *Pseudomonas syringae*, were practiced as foliar sprays and seed coatings, respectively, against collar rot of chickpea caused by *Sclerotium rolfsii*, with the combination of vermicompost substitution. When compared to controls, vermicompost substitutes reduced chickpea mortality significantly, but inhibition was much more efficient for treatments that included pre inoculation with nonconventional pesticides as foliar sprays against pathogen [47]. Vermicompost applications suppressed

Industry sludge type	Earthworm species used	Physico-chemical properties and heavy metals reduction	References
Sewage sludge derived biochar	*Eisenia fetida*	Biochar injected before composting lowered *E. fetida*'s bioavailability of Cd and Zn. Except for higher Cr concentrations, the biochar-added vermicomposts had good fertilizing capabilities.	Malińska et al. [50]
Municipal sludge mixed with cow dung	*E. fetida*	Cr, Cu, Ni, and Pb all the metal compounds were reduced after vermicomposting.	Srivastava et al. [51]

Table 6.
Effect of different types of earthworm species on heavy metal reductions of industrial sludge.

the tomato late blight caused by *Phytophthora brassicae*, *Phytophthora nicotianae*, and tomato Fusarium wilt produced by *Fusarium lycopersici*, as described by. Earthworm has stimulatory effect on soil microbial activities thus it suppressed the plant diseases more potentially than aerobic compost. There is a lot of research on the suppression effect of organic matter amendments in soils, with gratifying levels of reduction in plant parasitic nematode infestations. There are few scholarly publications on the suppressing effect of solid vermicomposts on numbers and outbreaks of plant parasitic nematodes relative to OM and thermophilic compost additives. Solid vermicompost applications for control of plant parasitic nematode populations have been studied [48]. Solid vermicomposts ranging from 2 to 8 kg ha^{-1} were applied to tomatoes, peppers, strawberries, and grapes in field treatments. They were able to suppress plant parasitic nematodes with great success. These researchers investigated the suppression capacity of plant parasitic nematodes in vermicomposts made from paper waste, food waste, and cattle manure under field circumstances and found considerable suppression.

5.6 Effect of vermicompost on bioremediation and detoxification of industrial wastes

Vermicompost has a greater importance in bioremediation and detoxification of industrial waste. Because of their robust metabolic system and the participation of earthworm gut bacteria and chloragocyte cells, earthworms have the potential to valorize and detoxification of heavy metals in industrial by-products. The majority of research found that vermicompost made from organic waste comprises greater concentrations of humic chemicals, which are important for plant growth [49]. Earthworm has a vast role in bioconversion of waste materials. Because of their robust metabolic system and participation of varied intestinal micro biota, enzymes, and chloragocyte cells that decrease hazardous forms to benign forms, earthworms have the ability to bio-convert and detoxify most heavy metals in industrial sludges (**Table 6**) [51].

6. Limitations of vermicompost

1. Vermicomposting is a time taking process. It requires almost 6-month for decomposing the organic wastes to prepare vermicompost.

2. In comparison to the traditional composting process, vermicompost requires higher maintenance.

3. Vermicompost may harbor pest and diseases as the temperature of vermicomposting pit have to be cool enough to support earthworm life.

7. Conclusion

Since vermicompost is organic in nature, it is not harmful for the environment. Vermicomposting process is also easy to operate and can be successfully prepared by unskilled small and marginal farmers. Amidst the environmental degradation and increasing food demand, vermicompost can be a solution. Although, its use alone in agriculture would not be able to meet the food demand but its use with chemical fertilizer through integrated manner can achieve sustainability in food production. The adoption rate of vermicompost is low and there is tendency of adopting vermicompost by female famers only. The potentiality of vermicompost is still not fully exploited yet. Hence, there is a need to appoint more extension worker to educate the farmers about vermicomposting and its benefits for achieving sustainability.

Author details

Priyanka Saha[1*], Anamika Barman[1] and Anurag Bera[2]

1 Division of Agronomy, ICAR-Indian Agricultural Research Institute, New Delhi, India

2 Department of Agronomy, Dr. Rajendra Prasad Central Agricultural University, Pusa, Bihar, India

*Address all correspondence to: priyankasaha9933@gmail.com

References

[1] Grappelli A, Tomati U, Galli E. Earthworm casting in plant propagation. Horticultural Science. 1985;**20**(5):874-876

[2] Kale RD, Bano K. Field trials with vermicompost (vee comp. E.83 UAS) an organic fertilizer. In: Dash MC, Senapati BK, Mishra PC, editors. Proceeding of National Seminar on Organic Waste Utilization Vermicompost. Part B: Verms and Vermicomposting. Burla, Orissa: Five Star Printing Press; 1986. pp. 151-156

[3] Bano K, Kale RD, Satyavathi GP. Vermicompost as fertilizer for ornamental plants. In: Rajagopal D, Kale RD, Bano K, editors. Proceedings of IV National Symposium on Soil Biology and Ecology (ISSBE). Bangalore: UAS; 1993. pp. 165-168

[4] Gomez-Brandon M, Dominguez J. Recycling of solid organic wastes through vermicomposting: Microbial community changes throughout the process and use of vermicompost as a soil amendment. Critical Reviews in Environmental Science and Technology. 2014;**44**(12):1289-1312

[5] Sharma K, Garg VK. Comparative analysis of vermicompost quality produced from rice straw and paper waste employing earthworm *Eisenia fetida* (Sav.). Bioresource Technology. 2018;**24**(8):7829-7836

[6] Sharma K, Garg VK. Vermi-modificaion of ruminant excreta using *Eisenia fetida*. Environmental Science and Pollution Research. 2017;**24**(24):19938-19945

[7] Soobhany N, Gunasee S, Rago YP, Joyram H, Raghoo P, Mohee R, et al. Spectroscopic, thermogravimetric and structural characterization analyses for comparing municipal solid waste composts and vermicomposts stability and maturity. Bioresource Technology. 2017;**263**:11-19

[8] Varghese SM, Prabha ML. Biochemical characterization of vermiwash and its effect on growth of *Capsicum frutescens*. Malaya Journal of Biosciences. 2014;**1**(2):86-91

[9] Chaoui HI, Zibilske LM, Ohno T. Effects of earthworms cast and compost on soil microbial activity and plant nutrient availability. Soil Biology and Biochemistry. 2003;**35**:295-302

[10] Brown GW, Moreno AG, Barois I, Fragoso C, Rojas P, Hernandez B, et al. Soil macrofauna in SE Mexican pasture and the effect of conversion from native to introduced pastures. Agriculture Ecosystems and Environment. 2004;**103**:313-327

[11] Pauli N, Barrios E, Conacher AJ, Oberthur T. Soil macrofauna in agricultural landscapes dominated by the Quesungual Slash and Mulch Agroforestry System, Western Honduras. Applied Soil Ecology. 2011;**47**:119-132

[12] Blouin M, Zully-Fodil Y, Pham-Thi A, Laffray D, Reversat G, Pando A, et al. Belowground organism activities affect plant aboveground phenotype, including plant tolerance to parasite. Ecology Letters. 2005;**8**:202-208

[13] Das MC, Saxena KG, Giri S. Vermitechnology for watershed reclamation, plant productivity and composting: A review in Indian context. International Journal of Ecology and Environmental Sciences. 2009;**35**:165-185

[14] Ansari AA, Ismail SA. Relamation of sodic soils through vermitechnology. Pakistan Journal of Agricultural Research. 2008;**21**:92-97

[15] Sailajakumari MS, Ushakumari K. Effect of vermicompost enriched with rock phosphate on the yield and uptake of nutrients in cowpea (*Vigna unguiculata*). Journal of Tropical Agriculture. 2002;**40**:27-30

[16] Arora VK, Singh CB, Sidhu AS, Thind SS. Irrigation, tillage and mulching effects on soybean yield and water productivity in relation to soil texture. Agricultural Water Management. 2011;**98**(4):563-568

[17] Parthasarthi K, Ranganathan LS. Supplementation of presumed vermicast with NPK enhances growth and yield in leguminous crops (*Vigna mungo* and *Arachis hypogaea*). Journal of Current Science. 2002;**2**:35-41

[18] Moledor S, Chalak A, Fabian M, Talhouk SN. Socioeconomic dynamics of vermicomposting systems in Lebanon. Journal of Agriculture, Food Systems, and Community Development. 2016;**6**(4):145-168

[19] Edwards CA, Burrows I. The potential of earthworm composts as plant growth media. In: Edwards CA, Neuhauser E, editors. Earthworms in Waste and Environmental Management. The Hague, the Netherlands: SPB Academic Press; 1988. pp. 21-32

[20] Nancarrow L, Taylor JH. The Worm Book: The Complete Guide to Gardening and Composting with Worms. Barkley, California: Wayback Machine Ten Speed Press; 1998. p. 4 [Archived: 18 March 2015]

[21] Sahariah B, Goswami L, Kim K, Bhattacharyya P, Sundar S. Metal remediation and biodegradation potential of earthworm species on municipal solid waste: A parallel analysis between Metaphireposthuma and *Eisenia fetida*. Bioresource Technology. 2015;**180**:230-236

[22] Kalantari S, Hatami S, Ardalan MM, Alikhani HA, Shorafa M. The effect of compost and vermicompost of yard leaf manure on growth of corn. African Journal of Agricultural Research. 2010;**5**:1317-1323

[23] Nada WM, Van Rensburg L, Claassens S. Communications in soil science and plant analysis effect of vermicompost on soil and plant properties of coal spoil in the Lusatian region (Eastern Germany). Communications in Soil Science and Plant Analysis. 2011;**42**(16):1945-1957

[24] Reddell P, Spain AV. Earthworms as vectors of viable propagules of mycorrhizal fungi. Soil Biology and Biochemistry. 1991;**23**(8):767-774

[25] Moradi H, Fahramand M, Sobhkhizi A, Adibian M, Noori M, Abdollahi S, et al. Effect of vermicompost on plant growth and its relationship with soil properties. International Journal of Farming and Allied Sciences. 2014;**3**(3):333-338

[26] Aksakal EL, Sari S, Angin I. Effects of vermicompost application on soil aggregation and certain physical properties. Land Degradation and Development. 2016;**27**(4):983-995

[27] Tharmaraj K, Ganesh P, Kolanjinathan K, Suresh KR, Anandan A. Influence of vermicompost and vermiwash on physico chemical properties of rice cultivated soil. Current Botany. 2011;**2**(3):18-21

[28] Chaoui I, Zibiliske M, Ohno T. Effects of earthworm casts and compost

on soil microbial activity and plant nutrient availability. Soil Biology and Biochemistry. 2003;**35**:295-302

[29] Edwards CA, Bohlen PJ. Biology and Ecology of Earthworms. 3rd ed. London: Chapman & Hall; 1996

[30] Kumar A. Decomposition of domestic waste by using composting worm *Eudrilus eugeniae* (Kinb.). In: Verms & Vermitechnology. New Delhi: APH publisher; 2005. p. 187

[31] Sharma S, Pradhan K, Satya S, Vasudevan P. Potentiality of earthworms for waste management and in other uses—A review. Journal of American Science. 2005;**1**(1):4-16

[32] Bhattacharjee G, Chaudhuri PS, Datta M. Response of paddy (Var. TRC-87-251) crop on amendment of the field with different levels of vermicompost. Asian Journal of Microbiology, Biotechnology & Environmental Sciences. 2001;**3**(3):191-196

[33] Roberts P, Jones DL, Edwards-Jones G. Yield and vitamin C content of tomatoes grown in vermicomposted wastes. Journal of the Science of Food and Agriculture. 2007;**87**(10):1957-1963

[34] Manivannan S, Balamurugan M, Parthasarathi K, Gunasekaran G, Ranganathan LS. Effect of vermicompost on soil fertility and crop productivity-beans (*Phaseolus vulgaris*). Journal of Environmental Biology. 2009;**30**(2):275-281

[35] Atiyeh RM, Edwards CA, Subler S, Metzger JD. Pig manure vermicompost as a component of a horticultural bedding plant medium: Effects on physicochemical properties and plant growth. Bioresource Technology. 2001;**78**(1):11-20

[36] Mahmoud IM, Mahmoud EK, Doaa IA. Effects of vermicompost and water treatment residuals on soil physical properties and wheat yield. International Agrophysics. 2015;**29**(2):157-164

[37] Islam MS, Hasan M, Rahman MM, Uddin MN, Kabir MH. Comparison between vermicompost and conventional aerobic compost produced from municipal organic solid waste used in *Amaranthus viridis* production. Journal of Environmental Science and Natural Resources. 2016;**9**(2):43-49

[38] Rasool R, Kukal SS, Hira GS. Soil organic carbon and physical properties as affected by long-term application of FYM and inorganic fertilizers in maize–wheat system. Soil and Tillage Research. 2008;**101**(1-2):31-36

[39] Nardi S, Morari F, Berti A, Tosoni M, Giardini L. Soil organic matter properties after 40 years of different use of organic and mineral fertilisers. European Journal of Agronomy. 2004;**21**(3):357-367

[40] Garcıa-Gil JC, Plaza C, Soler-Rovira P, Polo A. Long-term effects of municipal solid waste compost application on soil enzyme activities and microbial biomass. Soil Biology and Biochemistry. 2000;**32**(13):1907-1913

[41] Arancon NQ, Edwards CA, Atiyeh R, Metzger JD. Effects of vermicomposts produced from food waste on the growth and yields of greenhouse peppers. Bioresource Technology. 2004;**93**(2):139-144

[42] Acevedo IC, Pire R. Effects of vermicompost as substrate amendment on the growth of papaya (*Carica papaya* L.). Interciencia. 2004;**29**(5):274-279

[43] Mahmud M, Abdullah R, Yaacob JS. Effect of vermicompost amendment on nutritional status of sandy loam

soil, growth performance, and yield of pineapple (*Ananas comosus* var. MD2) under field conditions. Agronomy. 2018;**8**(9):183

[44] Edwards CA. Historical overview of vermicomposting. Biocycle. 1995;**36**(6):56-58

[45] Tomati U, Galli E, Grappelli A, Di Lena G. Effect of earthworm casts on protein synthesis in radish (*Raphanus sativum*) and lettuce (*Lactuga sativa*) seedlings. Biology and Fertility of Soils. 1990;**9**(4):288-289

[46] Patriquin DG, Baines D, Abboud A. Diseases, pests and soil fertility. In: Soil Management in Sustainable Agriculture. Wye, UK: Wye College Press; 1995. pp. 161-174

[47] Sahni S, Sarma BK, Singh KP. Management of *Sclerotium rolfsii* with integration of non-conventional chemicals, vermicompost and *Pseudomonas syringae*. World Journal of Microbiology and Biotechnology. 2008;**24**(4):517-522

[48] Arancon NQ, Galvis P, Edwards C, Yardim E. The trophic diversity of nematode communities in soils treated with vermicompost: The 7th International Symposium on Earthworm Ecology·Cardiff·Wales 2002. Pedobiologia. 2003;**47**(5-6):736-740

[49] Bhat SA, Singh S, Singh J, Kumar S, Vig AP. Bioremediation and detoxification of industrial wastes by earthworms: Vermicompost as powerful crop nutrient in sustainable agriculture. Bioresource Technology. 2018;**252**:172-179

[50] Malińska K, Golańska M, Caceres R, Rorat A, Weisser P, Ślęzak E. Biochar amendment for integrated composting and vermicomposting of sewage sludge—The effect of biochar on the activity of *Eisenia fetida* and the obtained vermicompost. Bioresource Technology. 2017;**225**:206-214

[51] Srivastava R, Kumar D, Gupta SK. Bioremediation of municipal sludge by vermitechnology and toxicity assessment by *Allium cepa*. Bioresource Technology. 2005;**96**(17):1867-1871

Section 2

Crop Management

Chapter 4

Sustainability-Based Review of Irrigation Schemes Performance for Sustainable Crop Production in Nigeria

Nura Jafar Shanono, Nura Yahaya Usman, Mu'azu Dantala Zakari, Habibu Ismail, Shehu Idris Umar, Sunusi Abubakar Amin and Nuraddeen Mukhtar Nasidi

Abstract

Irrigated agriculture has been identified as an important practice to achieving food security and socio-economic development in the face of rapid population growth and climatic uncertainties. In northern Nigeria, irrigation has long been identified as the key to achieving the much-desired increase in food production to meet the ever-increasing population. However, the existing irrigation schemes encountered several challenges coming from different dimensions including economic, social, environmental, institutional and technological. To attain sustainable crop production, this paper attempts to uncover the underline challenges confronting irrigation schemes in northern Nigeria that cut across sustainability pillars. The findings revealed that irrigation schemes contributed immensely toward achieving food security and improving the wellbeing of rural dwellers. However, the huge investment in large- and medium-scale irrigation schemes have resulted in massive economic losses. This could be attributed to their under-utilization, poor management and abandonment although few ones are performing remarkably well. The study recommends the need to adopt new water allocation and application methods that can improve water use efficiency, users-managers join approach (participatory), effective and competent institutions which include improved monitoring, evaluation and surveillance systems, frequent policy review to suit the situation, law enforcement, and timely sensitization and awareness campaigns.

Keywords: irrigation scheme, management, Nigeria, sustainable crop production, sustainability pillars

1. Introduction

Agriculture is an important sector in the economic development, poverty alleviation and in addressing food insecurity of many countries including Nigeria [1]. The problem of food insecurity represents the biggest crisis of the 21st century worldwide especially with the ongoing challenges posed by the notorious Corona Virus (COVID-19). The main point of concern here is that the impact of food insecurity is spreading from the developing to the developed countries of the world. According to the FAO report of 2018, about 821 million people do not have enough food, 2 billion people suffer from malnutrition and the numbers are rising at a high rate in both Africa and Asia [2]. Nigeria is not exceptional as its population is increasing at an alarming rate and this has glaringly highlighted the need for more food production to meet up and sustain the population demand. For example, the level of food insecurity in the rural areas of Nigeria is reportedly disturbing as it affected about 84% and 56% of the communities in northern and southern parts of the country respectively [3].

Nigeria relies mostly on the importation of agricultural products as about 31 and 23% of the total food demands were imported in 2011 and 2012 respectively [4]. For example, about 8 million metric tons of Rice and 5.6 million tons of Wheat were imported in 2019 to feed its growing population despite its production potential in agriculture [5]. Nigeria imported more than 10 million metric tons of Rice between 2010 and 2014 [6]. It has been suggested that the only way out of food insecurity and poverty is to remarkably attain a sustainable crop production in the country [7]. To improve agricultural productivity in the country, irrigation farming along with the use of improved seeds, fertilizers, mechanized and smart farming as well as other relevant and modern farming technologies is the best alternative option. This will help in reducing the level of hunger, poverty and malnutrition [8]. Therefore, irrigation can be regarded as a powerful factor in increasing crop productivity, more stable incomes and providing employment and increasing prospects for multiple cropping and crop diversification [9]. In the specific context of agriculture, sustainable irrigation strategies need to allow for increased and sustainable crop production to meet the ever-increasing food demands, while preserving natural resources [10, 11]. Moreover, irrigation farming allows farmers to produce all year round thereby resulting in higher agricultural outputs and improved farmer's income. According to [12], the objective of irrigation practice is to achieve the economical use of available water and ensure equity for distribution over time and space. In addition, the success of any irrigation project depends on the proper functionality of water conveyance and distribution systems. Unfortunately, many irrigation schemes in northern Nigeria are performing far below their potentials due to poor management by both relevant governmental agencies and farmers [13].

It was observed that improvement of the performance of the existing irrigation schemes is one of the possible approaches to water conservation, particularly in dry-land areas like northern Nigeria [14]. The term sustainability in irrigation is often characterized through indicators that express the performance of an irrigation scheme not only in terms of its ability to deliver the required irrigation water but also on economic viability, social wellbeing, environmental health, institution arrangement and technological advances. Thus, sustainable irrigated agriculture is said to be attained if irrigation practices do not lead to the depletion of either natural or human resources [11]. To meet the Nigerian population demands of food and fiber, there is a need to employ the concept of sustainability to further improve irrigated agriculture, thereby achieving sustainable food production, processing and value addition [15]. This can be achieved if all the causes and effects of many problems that have been lingering are diagnosed, propose solutions and the suggested solutions are implemented and put into practice. This review was therefore aimed to disclose the current status of the available irrigation schemes in northern Nigeria for their sustainability and functionality. The study will provide inside into what has been going on with regard to maintenance, utilization and level of crop productions in the irrigation schemes from sustainability point of view.

2. The state of irrigation schemes in northern Nigeria

This section presents the review on the state of irrigation schemes in northern Nigeria and sustainability pillars were used to guide this review. The aim was to get more insight into the global irrigation scheme operations and maintenance practices with the main focus on northern Nigeria. This review yielded a schematic overview of irrigation scheme management using sustainability pillars. This is to evaluate the present state of functionality and the level of impact on the lives of people. **Figure 1** shows the map of Nigeria showing the northern part, which constituted three regions (North West, North East and North Central), and southern part, which constituted three regions (South West, South East and South South). The irrigation scheme management sustainability-based review was conducted and restricted to the northern part of the country. The history of irrigation practices in the northern Nigeria begun since when it was realized that the region is characterized with low rainfall and high rate of evaporation which make it either arid or semi-arid regions in addition to abandon arable lands. These made the previous governments of the regions to construct several water storage infrastructures (dams and canals) for irrigation practices. These resulted to several major irrigation schemes available in northern Nigeria most of which are intended to stimulate and facilitate the sustainable food production in the country.

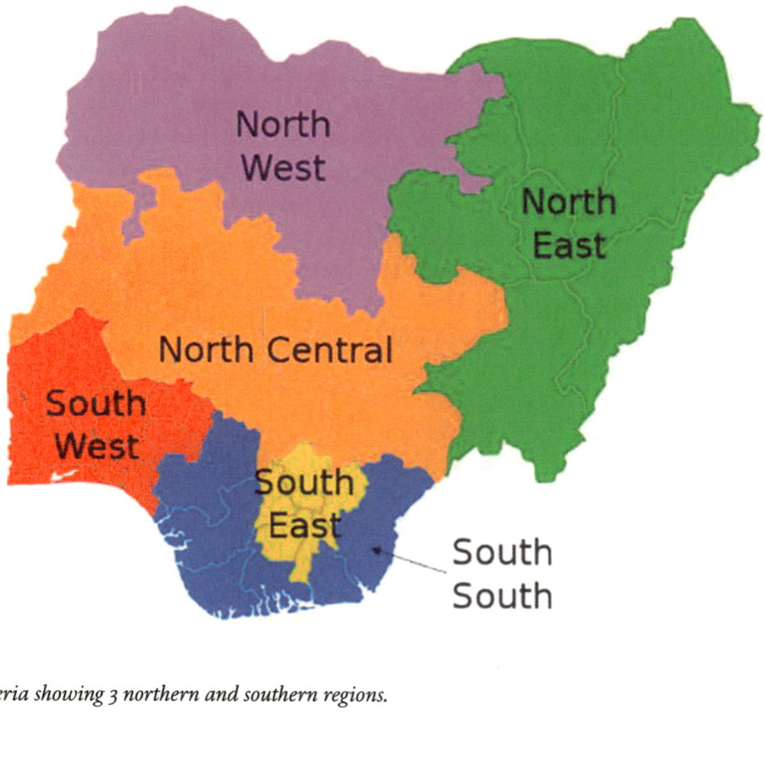

Figure 1.
Map of Nigeria showing 3 northern and southern regions.

2.1 Economic aspect of irrigation schemes sustainability

The economic productivity of several human endeavors depends largely on access to water resources [16]. Although about 24% of the global land area suffering from severe water scarcity [17] and 35% of the global population living in areas affected by water shortages [18], the economic development often occurs at the cost of over-exploitation of water resources [19]. Agriculture is a major performer in the human appropriation of the limited water resources as about 70% of the global freshwater is consumed by this sector. After abandonment for about four decades due to abundant petroleum resources, the agricultural sector in Nigeria is gradually occupying a dominant position in the development of the national and rural economy. The sector provides not only food but also serves as the major source of employment to the teeming population of Nigeria. The agricultural sector provides jobs to about three-quarters of the Nigerian working population [20]. Farmers are usually less busy on the farm during the dry season, putting into account the rainy (May to October) and the dry (November to April) seasons of Nigeria. Hence, the provision of irrigation facilities that offer the opportunity for all-year-round farming can serve as an alternative source of employment and an additional gain to the Nigerian economy [21].

Recently, drastic agricultural reforms (closure of land borders and banning of importation of major agricultural food products among others) have been made in Nigeria resulting in a sharp increase in crop production which significantly reduce food importation and jobs were created [22, 23]. Agriculture is one of the main economic sectors in Nigeria employing about 60% of the population of the country [23]. This scenario is in line with other developing countries that agriculture provides the leading source of employment. Thus, increasing agricultural productivity is critical to economic growth, development, and the nation's Gross Domestic Products (GDP). One important way to increase agricultural productivity is through the introduction of improved agricultural technologies and management systems.

In Nigeria, post-project evaluations of the majority of the irrigation schemes revealed that their economic performances are low compared to pre-project predictions [14, 15]. Such undesirable outcomes are a result of the fact that social and environmental concerns of these schemes were not incorporated in the analysis. The participation issue presents the usefulness of water users' involvement in the maintenance and sustainability of the irrigation schemes which further improve economic benefit [24]. One of the possible causes of the decline in food production is an inefficient allocation of resources in the agricultural production potentials [25]. For example, Land, labour, capital and water resources are inefficiently allocated thereby leading to a decrease in their productivity. To further improve the economic status of the rural dwellers as well as to attain food security and national growth, irrigation schemes need to be revitalized to increase food and cash crops production [26].

The economic welfare of a country and its ecosystem health is directly linked to water stress and the rate of water depletion [24]. Evidently, the Kano River Irrigation Project (KRIP) has played an important role in discouraging migration from rural to urban centers and alleviating the employment problems of its immediate community [20]. Similarly, there has been a sharp difference from the dry season farmers' income in Bauchi State when compared with rain-fed farmers for the same kind of farm produce. The dry season (irrigation) farmers get more profit than rain-fed farmers counterparts. This is not unconnected with the high demand for fresh irrigated crops during the dry season [27]. Moreover, a study on the Socio-economic impact of an irrigation project in Taraba State reported similar findings among the farmers in respect of economic gain. Project on the beneficiary of the Fadama II project in Kaduna State indicates an increase in the net farm income of the beneficiary farmers [28]. Hence, creating a more efficient irrigation water management approach has the potential to substantially increase agricultural production, farmers' incomes, and create employment opportunities.

2.2 Social aspect of irrigation schemes sustainability

Naturally, human beings have inherent needs that they are aiming to satisfy. These needs describe in-born requirements that need to be satisfied for an individual to remain healthy – physically, emotionally and mentally [29]. The management of open-access resources such as irrigation water involved numerous stakeholders with diverse interests which posed a unique challenge to the managers. These interests are the factors that affect individuals' ethical practices including propensity to compliant or unlawful activities. Ethical awareness is the ability of an individual to identify his deliberate action and understand what consequences that action might cause to others. Thus, for an individual to make an ethically accepted action depends on a person's moral awareness, motives and the benefits that individual is expected to gain. This depends largely on value-related factors such as culture, knowledge, religious beliefs, public trust and social wellbeing [30]. One of the problems that devastate irrigation water users' social wellbeing is water scarcity which leads to poor crop production. Water scarcity represents a multidimensional state of human social deprivation characterized by a lack of access to affordable and safe water to satisfy societal needs or a condition in which these needs are met at the expense of the environment [10]. The mission for sustainable natural resource utilization is an essential part of the ongoing 2030 agenda for sustainable development goals (SDGs). It is one of the 169 agreed targets being aimed at monitoring and assessing the level of sustainability with which resources, such as irrigation water are being managed and utilized [31].

This creates a challenge of ensuring societal wellbeing through the supply of human basic needs including food security, income to the rural dwellers and national GDP of which water (through irrigation) plays an essential role [32, 33]. Thus, irrigation contributed immensely to the provision of a wide range of socio-economic benefits on which the wellbeing of society is based [34]. However, irrigation water is subjected to several challenges including climate change, poor management, chemical, wastage, overexploitation and other human-related influences [35–38].

The social setting can be a social group, a community, town, region or a nation, thus, any change that occur either as ideas, norms, values, roles and social habits can be referred to as social change. When alteration occurs in the rural social system, it is termed as rural social change, and such a change could be in all attributes of a societal unit such as number, quality and importance. Different changes come to the notice of the rural population of the developing countries, including Nigeria. For example, the introduction of large scale irrigation projects, use of the machine in farming practices, application of agrochemicals to control weeds, pests, diseases and increase and sustain the fertility of the soil lead to the transformation of sustainable agriculture and hence, the wellbeing of that society [20]. The attainment of a sustainable agricultural production system is becoming a major concern of agricultural researchers and policymakers all over the world [39]. Implementation of sustainable development, therefore, requires integrated policy, planning and social learning process. Irrigation practices provide employment and stabilization to the rural population and undoubtedly provide major social benefits. A typical example is how the Kano River Irrigation

Project (KRIP) played an important role in limiting rural–urban migration by creating jobs for the rural dwellers [20].

2.3 Environmental aspect of irrigation schemes sustainability

Environmental impact refers to any change in the environment or in its components that may affect human health, flora, fauna, natural and cultural heritage as well as other physical structures, social, economic or cultural conditions [40]. For example, the challenges facing the irrigation sector in Nigeria is not only to attain food security and eradicate poverty among rural dwellers but also to ensure a healthy environment. Inappropriate management of irrigation schemes might lead to environmental problems such as high water tables, poor drainage, salinization and pollution [13]. The majority of irrigation schemes in northern Nigeria are characterized by environmental degradation such as salinity, waterlogging and declining groundwater resources which could adversely affect future demand for water [41]. Both quantity and quality aspects of water are important as these jointly affect the success of irrigation schemes and environmental sustainability [42]. Thus, in the process to establish any socio-economic projects such as irrigation schemes, there is a need to ensure long-term maintenance of valued environmental resources in an evolving human influence [43]. Studies revealed that the majority of the economic development of the developing countries often occur at the expense of overexploitation of water resources which ultimately leads to ecosystem degradation [19].

Even though the extent is different, several environmental-related problems including soil erosion, aquatic weeds infestation, sedimentation, infrastructural deterioration and overgrazing are observed in many irrigation schemes in Nigeria [43]. For example, despite the functioning of the Kano River Irrigation Project (KRIP), there was a serious decline in hectares of land due to environmental-related issues such as waterlogging, salt accumulation (salinity, sodicity, saline-sodic) and reduced fertility [20]. There is a gradually building up of salinity problems in KRIP, even though the threat from salinity is not alarming yet. This problem of salinity has been reportedly alleged to continue to increase as long as irrigation is practiced unless preventive and corrective measures are put in place [44]. Generally, irrigation schemes design, operation and management should seek to maximize not only crop productivity and economic and social gains but also to ensure environmental stability and health as shown on **Figure 2** [46]. Thus, irrigation scheme designs should consider using new technologies that ensure water allocation and application efficiencies such as micro irrigation methods (sprinklers and drip). In addition, in situ soil and water conservation methods such as mulch practices and deficit irrigation can significantly improve the overall ecosystem health.

2.4 Technological aspect of irrigation schemes sustainability

The development and improvement achieved so far in irrigation technologies are key to addressing the challenges of low agricultural productivity [47]. Availability and access to irrigation water and smart agricultural technologies were considered essential for crop production [48]. For instance, the success of the green revolution in Asia was achieved through the rapid expansion of irrigation areas with availability and access to new technologies including the development of high yielding varieties, fertilizers, micro irrigation techniques, tube-wells and water extraction mechanisms [49]. For example, technological advance provides irrigation sector with methods of optimizing water usage using variety of solutions based on sensor networks, microcontrollers and machine learning or fuzzy logic [45]. These methods have been in use to evaluate and predict optimum water required for irrigation. Such a smart irrigation is a systems made up of solar power station, networking infrastructure and water management and control stations (water storage, sprinkle or drip lines, water pumps, soil moisture sensors and micro-controller unit). In smart irrigation systems, the system components are commonly coupled using the Internet of Everything (IoE) approach as schematically summarized and shown in **Figure 3**. The use of such irrigation technology exerted a positive and significant impact on sustainable crop production and food security in Nigeria specifically [48].

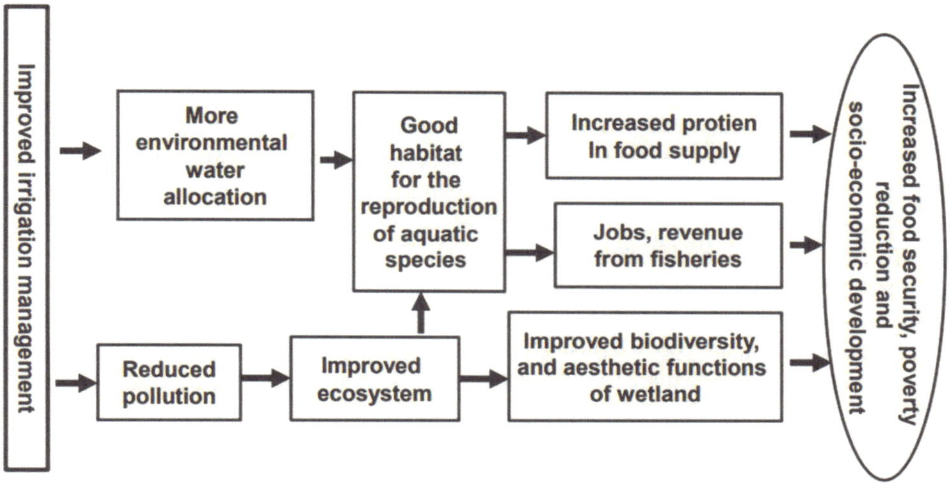

Figure 2.
Framework for sustainable irrigation scheme design, operation and management [45].

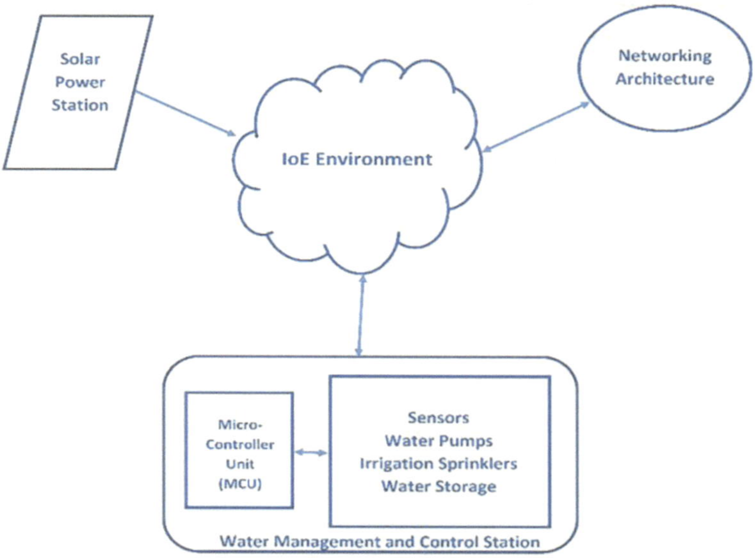

Figure 3.
Technological advances for coupling smart irrigation components using IoE [45].

Agricultural activities in Jibia Irrigation Project (JIP), Katsina State, depends mainly on power supply from diesel generators and electricity from the national grid to supply water to the farmlands. This has slowed down the pace of irrigation development in the area. The full exploitation of the agricultural potential of JIP and that of Nigeria in general, requires the exploitation of our vast renewable energy sources to provide the needed power. Also, an alternative way to the high power demand of the operation of JIP and the likes is water conservation practices using different types of mulch. KRIP being one of the major irrigated agriculture in the northern Nigeria reported that, the majority of the farmers lack the technical know-how on water conservation and it is based on this that the researchers recommended the need to create awareness to the farmers where major irrigation is taking place. This will assist in achieving water conservation and management strategies in order to effectively and efficiently utilize the limited available water resources.

Irrigation schemes in Nigeria such as Watari Irrigation Project (WIP), Barwa-Minjibir Irrigation Scheme (BMIS), Tomas Irrigation Project (TIP) and Kpong Irrigation Project (KIP) in Niger Delta Basin Development Authority (NDBDA), had to abandon farming activities due to poor water-sharing techniques [50]. Thus, farmers should be fully conversant with irrigation technologies through agricultural machinery and credit facilities. Moreover, farm inputs such as fertilizer, seed, chemical and other materials needed by farmers should be made available to the farmers. Infrastructural decay is also another problem that has been affecting the success of irrigated agriculture in northern Nigeria. About 30% of water structures at WIP were found to be damaged and malfunctioning [51]. Similarly, the conveyance structures were silted and infested by weeds which significantly reduced the carrying capacity of the canal. Also, about 8% of the irrigable area downstream was abandoned due to inadequate supply of water. There is an increased occurrence in soil salinity and sodicity issues within the WIP due to a poor drainage system [52].

In this regard, the majority of the irrigation schemes in northern Nigeria such as Sokoto Rima, Watari, Jibiya and Tomas are operated far below their design capacity due to lack of adoption of improved equipment and poor maintenance [53]. In addition, a larger proportion of the currently used irrigation equipment was purchased during the inception of the projects (the 1970s to 1980s) without replacement. Thus, there is a need to conduct empirical studies in Nigerian irrigation schemes to assess the following;

a. How farmers demand irrigation technologies.

b. How their zeal and willingness to adopt improved irrigation technologies are affected by specific agro-ecological and socio-economic characteristics.

c. How their adoption of such improved irrigation technologies may be hampered by poorly functioning markets.

2.5 Institutional aspect of irrigation schemes sustainability

Institutions are the political, social or business organizations (public or private) that are involved in policy-making and implementation. While institutional sustainability is the continuation of the benefit flows to the users/clients/owners/employees or the general public with or without the programmes or organizations that stimulated them in the first place [54]. Institutional performance is considered as one of the yardsticks with which the performance of developmental institutions such as irrigation schemes can be evaluated. At the end of the 20th century, the increasing role and relevance of social and institutional structures in connection with the whole field of contemporary environmental management are gaining prominence. Currently, institutional mandates constituted social well-being, economic gain as well as environmental health. Such a sustainability-based management strategy has gained more attention all over the world as this form an important developmental strategy as enclosed in the ongoing sustainable development goals (SDGs). The main aim of such a strategy was to effectively and sustainably manage and utilize the limited available natural resources [55]. For example, in the irrigation management sector, this approach has in recent years been employed to shift irrigation management toward a community-based by sharing power with multiple sets of other institutions stakeholders [56]. This requires every stakeholder involved in all levels of irrigation management to collectively take responsibility for managing the affairs of the schemes.

The small-scale private irrigation schemes (SPRI) sector in Nigeria is supported by a range of private agents, including irrigation technology service providers, NGOs, water user associations (WUAs) as well as public institutions such as the National Fadama Development Project (NFDP), the Agricultural Development Project (ADP), the State Irrigation Department (SID), river basin development authorities and state and federal government ministries [56]. A study carried in 1972 led to the institution of three models of public irrigation schemes; namely the Bakolori Scheme, the Chad Basin scheme, and Kano River Irrigation Project, subsequently additional eleven more River Basin Development Authorities (RBDAs) were added across the country after the success of the pilot schemes in 1976 [57].

The Nigerian government does not only own, operate and maintain irrigation schemes, but provides agro-support services such as land preparation, seeds, fertilizers, chemicals, and assists in marketing the produce. The reforms in water institutions such as Participatory Irrigation Management (PIM) systems were formulated and implemented to achieve effective operation of the schemes, equitable distribution of irrigation water among farmers, high crop productivity and food security among others [58].

In the Hadeja-Jama'are river irrigation project, the utilization of the project is just 50% while the Zobe dam in Dutsin-Ma, Katsina which was constructed 40 years ago, currently has few irrigation activities as the scheme is not formally developed. Also, at the Bakolori irrigation dam in Zamfara State, under the Sokoto Rima Water Project, the area cultivated is not commensurate with the amount of water in the dam [57]. For instance, at the end of the 1999/2000 irrigation season, out of the 100,300 ha developed only 35,000 ha were irrigated giving a pathetic 35% capacity utilization. Most of the irrigation schemes that the government has invested in are either underutilized for irrigation or abandoned irrigation schemes like the Hadeja-Jama'are river project, the utilization is 50% while the Zobe dam in Dutsin-Ma in Katsina, which was constructed 40 years ago, currently has little irrigation activities [52]. Cases in points that highlighted the danger of poor irrigation management institutional performance are the findings by [53]. More than 29% of the farmers of Tomas Irrigation Project (TIP) expressed unhappiness with the water allocation method currently used and about 55% of water users hold the opinion that irrigation scheme management, operation and maintenance is an exclusive responsibility of the government. In addition to poor water management, infrastructural decay and stakeholders' conflict as the major problems affecting the scheme.

3. Conclusion

This chapter presented a review of the operation and management of irrigation schemes in the northern part of Nigeria. The main aim of this study was to identify and relate key values operating in the northern Nigerian irrigation sector from a sustainability point. This includes identifying the causes and effects of impending problems and hence, suggesting ways forward to achieve sustainable food security at the face of the ever-increasing population in the country. The major motivation factor for this review was the fact that studies on assessing the performance of systems such as irrigation schemes using the concept of sustainability are gradually gaining popularity and growing at a high rate in recent years. The appraisal revealed that several impediments have been hindering the performance of irrigation practice in Nigeria which includes inconsistent government policies, low awareness and lack of technical know-how among the farmers on irrigation farming system, and untimely financial intervention.

Another insight of interest gained from the work was that the huge investment in both large- and medium-scale irrigation schemes in the northern part of Nigeria have been resulted in irrecoverable losses due to quite several issues. Some of these problems comprise of under-utilization of water resources, poor management, infrastructural decay and abandonment. Generally, studies revealed that irrigation schemes in northern Nigeria performed far below expectations with approximately 65% capacity utilization. About half (50%) of the farmers express unhappiness, dismay and loss confidence with the way irrigation schemes are operated. Water managers blame farmers to lack enthusiasm toward abiding by the set rule and regulations governing irrigation schemes. In addition, study by [50] revealed that about 45% of the farmers do not participate in the maintenance of the irrigation schemes which further exacerbate the problems. However, some irrigation projects are performing relatively very well. The income and the standard of living of the farmers around well-performing irrigation projects were observed to improve significantly compared to poor performing ones.

Thus, there is a need to holistically improve the general operational performance of the existing irrigation schemes in northern Nigeria through the following;

- To encourage participatory irrigation management (users-managers join management).

- To ensure effective and competent institutions through improving the monitoring and evaluation mechanisms (surveillance systems, frequent policy review and alterations to suit the situation, legal proceedings and law enforcement, and timely sensitization and awareness campaigns).

- Functions of water regulatory institutions should be streamlined with each institution given specific and defined roles to enhance efficiency in irrigation water resource management and this should be organized using the sustainability pillars as schematically shown in **Figure 1**.

- A research effort using a sustainability-based approach is also required to further identify the causes and effects of problems that have been hampering the performance of irrigation schemes in northern Nigeria.

- There is a need to adopt new water allocation and application methods that can improve water use efficiency.

Acknowledgements

On behalf of the entire authors, I would like to thank the management of Bayero University Kano, Nigeria for the opportunity to conduct this study though Directorate of Research, Innovation and Partnerships (DRIP), Institution Based Research (IBR), a component of the Tertiary Education Trust Fund (TETFund).

Conflict of interest

The authors declare no conflict of interest.

Author details

Nura Jafar Shanono[1*], Nura Yahaya Usman[1], Mu'azu Dantala Zakari[1], Habibu Ismail[2], Shehu Idris Umar[3], Sunusi Abubakar Amin[4] and Nuraddeen Mukhtar Nasidi[1]

1 Department of Agricultural and Environmental Engineering, Bayero University Kano, Nigeria

2 Department of Agricultural and Bioresources Engineering, Ahmadu Bello University, Nigeria

3 Department of Hospitality and Tourism Management, Federal University Wukari, Nigeria

4 Department of Agricultural and Bioresources Engineering, Abubakar Tafawa Balewa University, Nigeria

*Address all correspondence to: njshanono.age@buk.edu.ng

References

[1] Hassan IYAMB, Yahaya R. Economic analysis of micro-drip irrigation using integrated Agricultural Research for Development (Iar4d) approach: The case of vegetable innovation platform in rural Nigeria. Agricultural Transformation in a Deregulated Economy: Prospects and Challanges. 2012:103-106

[2] Bayero MT. Assessing the Sustainability of Drainage System in Irrigated Agricultural Land: A Case Study of Kano River Irrigation Scheme in Nigeria Defended. Pan-african university institute for water and energy sciences (including climate change); 2019. p. 1-133. Available from: http://repository.pauwes-cop.net/handle/1/348

[3] Akinyele IO. Ensuring Food and Nutrition Security in Rural Nigeria: An Assessment of the Challenges , Information Needs, and Analytical Capacity. International Food Policy Research Institute: Washington D.C; 2009

[4] Astou D. Food imports as a hindrance to food security and sustainable development: The cases of Nigeria and Senegal. City University of New York. 2015. p. 1-61. Available from: https://academicworks.cuny.edu/cgi/viewcontent.cgi?article=1554&context=cc_etds_theses

[5] Beillard MJ, Nzeka UM. Assessments of commodity and trade issues: Nigeria grain and feed annual 2019 Nigeria's imports of wheat and rice to rise. U. S Government; 2019. https://apps.fas.usda.gov/newgainapi/api/report/downloadreportbyfilename?filename=Grain%20and%20Feed%20Annual_Lagos_Nigeria_5-6-2019.pdf. GAIN Report Number: NG-19002

[6] Yusuf WA, Yusuf SA, Adesope AAA, Adebayo OZ. Determinants of Rice import demand in Nigeria. Journal of Applications. 2020;**24**(May):923-931. DOI: 10.4314/jasem.v24i5.30

[7] Xiea TH, Youa HL. Invest in small-scale irrigated agriculture: A national assessment on potential to expand small-scale irrigation in Nigeria. Agricultural Water Management. 2017;**193**:251-264

[8] Dauda BF, Asiribo TO, Akinbode OE, Saka SO, Salahu JO. An assessment of the roles of irrigation farming in the millennium development goals. African Journal of Agriculture Research. 2009;**4**(5):445-450

[9] C D, Joseph JJN, Maurice M. Profitability assessment of irrigated crop production among small-scale farmers in Gombe state, Nigeria. Direct Research Journal of Agriculture F. 2019;**7**(7):166-172

[10] Rosa L, Chiarelli DD, Rulli MC, Angelo JD, Odorico D. Global agricultural economic water scarcity. Science Advance. 2020;6(8). DOI: 10.1126/sciadv.aaz6031

[11] Borsato E, Rosa L, Marinello F, Tarolli P. Weak and strong sustainability of irrigation: A framework for irrigation practices under limited water availability. Front Sustainability Food System. 2020;**4**(February):1-16

[12] Nasidi NM, Shanono NJ. Performance evaluation of water conveyance system at Watari irrigation project (WIP). Proceedings iSTEAMS Multidisciplinary Cross-Border Conference. 2016;**3**:105-110

[13] Jibril AAY, Saidu GM. Performance evaluation of Badeggi irrigation scheme, Niger state Nigeria, using efficiency techniques. Scholarly Journal of Science Research Essay. 2017;**6**(July):42-47

[14] Shanono NJ, Nasidi NM, Zakari MD, and Bello. Assessment of Field Channels Performance At Watari Irrigation Project Kano, Nigeria. In: Proceedings of the Nigerian Institute of Agricultural Engineering (NIAE). Bauchi State, Nigeria. 2012. pp. 144-150.

[15] Mohammed U and Ali MS. Socio economic challenges of irrigation farming along river Yobe (a case study of Yobe state). 2021; **15**(4):1-7. DOI: 10.9790/2402-1504010107

[16] Davis KF, Rosa L, Carr JA, Chiarell D, Angel GKMJD, Gephart J, et al. Reviews of geophysics the global food-energy-water nexus. Advances of Earth Space Science. 2018;56(3):456-531. DOI: 10.1029/2017RG000591

[17] Alcamo J et al. Global estimates of water withdrawals and availability under current and future 'business-as- usual' conditions. Hydrological Sciences Journal. 2010;**48**(3):339-348

[18] Rockström J et al. Ecohydrology bearings — Invited commentary the unfolding water drama in the Anthropocene: Towards a resilience-based perspective on water for global sustainability. Ecohydrology Bearings. 2014;**1261**(7):1249-1261

[19] Savenije H, Van Der Zaag P, Delft IHE. Water as an economic good and demand management paradigms with pitfalls. International Water Resoures Association. 2002;**27**(1):98-104

[20] Haruna SK. Impact of Participatory Irrigation Management (PIM) on the Livelihood of Water Users in Kano River Irrigation Project (KRIP), Nigeria: Rural Sociology, Ahmadu Bello University; 2015. Available from: http://kubanni.abu.edu.ng/jspui/bitstream/123456789/8363/1/impact%20of%20participatory%20irrigation%20management%20%28pim%29%20on%20the%20livelihood%20of%20water%20users%20in%20kano%20river%20irrigation%20project.pdf

[21] Hassan A, Adewumi MO, and Falola A. An assessment of the irrigation scheme on registered Rice farmers of the upper Benue Rice basin development Authority in Dadin Kowa, Gombe state, Nigeria. 2015;**4**(1):1-24. DOI: 10.7828/jmds.v4i1.843

[22] Kolawole AO, Oluwatusin FM, Ajiboye A, Aturamu OA, Abdu-Raheem KA, Akokoh FE. Poverty status analysis of irrigation farming households in Nigeria. World Rural Observations. 2020;**12**(2):15-26. DOI: 10.7537/marswro120220.02

[23] FMWRN. National irrigation and drainage policy and strategy. Abuja-Nigeria. 2015. p. 1-26. Available from: https://silo.tips/download/national-irrigation-policy-and-strategy-for-nigeria-part-1-the-need-for-a-policy

[24] Borsato E, Martello M, and Marinello F. Environmental and Economic Sustainability Assessment for Two Di Ff Erent Sprinkler and a Drip Irrigation Systems: A Case Study on Maize Cropping. 2019;9(187):1-15. DOI: 10.3390/agriculture9090187.

[25] Ohikere AJZ, Ejeh AF. Impact of small scale irrigation technologies on crop production by fadama users in Kogi state, Nigeria. Advances in Applied Science Research. 2012;**3**(2):854-861

[26] Ahmed E, Oyebode A, Igbadun MA, Oiganji HE. Assessment of tomato farmers' irrigation practice in pampaida millennium village, ikara local government area, Kaduna state, Nigeria. FUDMA Journal of Science. 2020;**4**(2):499-509

[27] Adama GJ, Jimoh DO, Otache MY. Optimization of irrigation water allocation framework based on genetic algorithm approach. Journal of Water Resource and Protection. 2020;**12**:316-329

[28] Abdullahi AS, Jahun BG, Sabo MU. Impact of irrigation project on fadama community of Bauchi state Nigeria. International Journal of Advances Engineering Science. 2016;**5**(1):20-30. Available from: https://www.academia.edu/27487403/impact_of_irrigation_project_on_fadama_community_of_bauchi_state_nigeria

[29] Missimer G, Robèrt K-H, Broman G. A strategic approach to social sustainability - part 1: Exploring the social system. Journal of Cleaner Production. 2016. DOI: 10.1016/j.jclepro.2016.03.170

[30] Treviño LK, Weaver GR, Reynolds SJ. Behavioral ethics in organizations: A review. Journal of Management. 2006;**32**(6):951-990

[31] Shanono NJ. Towards a more human-centered irrigation water management- a review. International Journal of Water Management Diploma. 2021;**1**(3):5-16. Available from: https://dergipark.org.tr/en/pub/ijwmd/issue/63570/943011

[32] Manju S, Sagar N. Renewable energy integrated desalination: A sustainable solution to overcome future fresh-water scarcity in India. Renewable and Sustainable Energy Reviews. 2017;**73**:594-609. DOI: 10.1016/j.rser.2017.01.164

[33] Hossain S et al. Identifying future research directions for biodiversity, ecosystem services and sustainability: Perspectives from early-career researchers. International Journal of Sustainable Development & World Ecology. 2018;**25**(3):249-261

[34] Wang M, Li J, Ho Y. Research articles published in water resources journals: A bibliometric analysis. Desalination and Water Treatment. 2011;**28**(4):353-365

[35] Zhang Y, Chen H, Lu J, and Zhang G. Detecting and Predicting the Topic Change of Knowledge- Based Systems: A Topic-Based Bibliometric Analysis from 1991 to 2016. 2016;10-19:10-19.

[36] Shanono NJ, Ndiritu J. A conceptual framework for assessing the impact of human behaviour on water resource systems performance. Algerian Journal of Engineering. 2020;**02**(35-44). DOI: 10.5281/zenodo.4400183

[37] Shanono NJ. Co-evolutionary dynamics of human behaviour and water resource systems performance: A socio-hydrological framework. Academic Letters. 2021;**1191**:1-6

[38] Nasidi NM, Shanono NJ, Zakari MD, Ibrahim A, Bello MM. Reclaiming salt-affected soil for the production of tomato at Barwa-Minjibir irrigation scheme, Kano. In: International Conference on Green Engineering for Sustainable Development, IC-GESD 2015. Kano Nigeria: Held at Bayero University; 2015

[39] Medugu N. Achieving Sustainable Agriculture in Nigeria: A Land-Use Policy Perspective. Tokyo: Industry & Cultural Integration Tour; 2006;10-19:10-19. Available from: http://eprints.utm.my/id/eprint/3538/1/idris.pdf.

[40] Ulsido MD, Demisse EA, Gebul MA, Bekelle AE. Environmental impacts

of small scale irrigation schemes : Evidence from Ethiopian Rift Valley Lake basins environmental impacts of small scale irrigation schemes: Evidence from Ethiopian Rift Valley Lake basins. Environmental Research, Engineering and Management. 2013;**1**(63):17-29

[41] Sobowale A, Tijani MN, Obayelu AE, Olatunji AS, Shah T. Livelihood analysis of smallholder irrigation farmers In Nigeria. Journal of Agriculture Science Environment. 2014;**14**:1-17

[42] Mohamme A, Ibrahim HA. Variabilty of irrigation water quality in Kano River irrigation project. JORIND. 2015;**13**(2). Available from: www.ajol.info/journals/jorind

[43] Purity N, Adaeze EE. Environmental sustainability and sustainable development in Nigeria : Environmental sustainability and sustainable development in Nigeria: Problems and prospects. International Journal of Academic Accounting, Finance Managment Research. 2020;**4**(1):6-11

[44] Maina MM et al. Soil salinity assessment of Kadawa irrigation of the Kano River irrigation project (KRIP). Journal of Food Agriculture Environment. 2012;**10**(October)

[45] Favour A, Misra S, Maskeliunas R, Damasevicius R, Kazanavicius E. Smart irrigation system for environment sustainability in Africa: An Internated of everything (IoE) approach. Mathematical Biosciences and Engineering. 2019;**16**(5):5490-5503 https://www.10.3934/mbe.2019273

[46] Fonteh MF. Guidelines for sustainable irrigation system design and management in sub-Saharan Africa. African Journal of Agricultural Research. 2017;**12**:1747-1755

[47] Ngango J, Hong S. Adoption of small-scale irrigation technologies and its impact on land productivity: Evidence from Rwanda. Journal of Integrative Agriculture. 2021;**20**(8):2302-2312

[48] Adebayo O, Bolarin O, Oyewale A, Kehinde O. Impact of irrigation technology use on crop yield, crop income and household food security in Nigeria: A treatment effect approach. Agriculture Food. 2018;**3**(June):154-171

[49] Madhusudan B, Narayanamoorthy A. Impact of irrigation on agricultural growth and poverty alleviation: Macro level analyses in India. Water Policy Research. 2004. p. 1-8. Available from: https://dlc.dlib.indiana.edu/dlc/bitstream/handle/10535/5130/TATA_2003_Bhattarai.pdf?sequence=1

[50] Shanono NJ, Bello M, Muntaqa I, Usman T. Stakeholders conflict and infrastructural decay In Nigerian irrigation schemes: A review. Nigeria Journal of Engineering, Science Technology Research. 2020;**6**(1):78-90

[51] Shanono NJ et al. Hydraulic infrastructures and Asseeement of Watari irrigation project, Kano. Journal of Engineering Technology. 2015;**10**(2):44-51

[52] Bashir et al. Appraisal and mapping of soil salinity and Sodicity problems In sector one of Watari irrigation scheme, Kano state. Nigeria Soil Journal of Social Science. 2020;**29**(2):54-60

[53] Shanono NJ et al. Socio-hydrological study of water users' perceptions on the management of irrigation schemes at tomas irrigation project, Kano, Nigeria. Nigeria Journal of Engineering, Science Technology. 2019;**5**(2):139-145

[54] Mugabi J, Kingdom W, Kayaga S. Evaluating the institutional sustainability

of an urban water utility: A conceptual framework and research directions. Elsevier. 2013;**3**:1-21

[55] Saravanan VS, Bhawan P. Institutionalising community-based watershed management in India: Elements of institutional sustainability. Water Science and Technology. 1990;**45**(11):113-124

[56] Takeshima H, Okoli SS, Rhoe V. Demand Characteristics for Small-Scale Private Irrigation Technologies: Knowledge Gaps in Nigeria. The Nigeria strategy support program (NSSP); 2010. NSSP Working Paper No. 0018. Available from: https://www.ifpri.org/publication/demand-characteristics-small-scale-private-irrigation-technologies

[57] Adelodun B, Choi K. A review of the evaluation of irrigation practice in Nigeria: Past, present and future prospects. African Journal of Agriculture Research. 2018;**13**(40):2087-2097

[58] Ahmad B et al. Impact of institutional features on the overall performance assessment of participatory irrigation management: Farmers' response from Pakistan. MDPI. 2020;**12**(497):1-13

Chapter 5

Crop Diversification an Effective Strategy for Sustainable Agriculture Development

Anamika Barman, Priyanka Saha, Shashank Patel and Anurag Bera

Abstract

Sustainable agricultural practices involve a variety of approaches. The most important approached for sustainable agriculture development is crop diversification. It allowing the farmers to employ biological cycles to minimize inputs, conserve the resource base, maximize yields and also reduce the risk due to ecological and environmental factors. It serves as an important opportunity to augment income and employment generation for rural communities. Crop diversification promotes the interaction of beneficial soil bacteria, interrupts the disease cycle, and reduces the quantity of weeds. Crop diversification boosts land-use efficiency and crop output by improving the physical and chemical qualities of soil. Crop diversification shows a lot of scope to alleviating the problems such as resurgence of insects-pests and weeds, soil degradation, environmental pollution, soil salinity, decline farm profit and climate change. Crop diversification through crop intensification system enhanced the net returns, B:C ratio, and overall system productivity of a farm. In order to achieve the benefits of crop diversification farmers are shifting from low value low yielding crops to high value high yielding crops. Thus, crop diversification has the sound capacity for achieving the goal of nutritional security, income growth, food security, employment generation and sustainable agriculture development.

Keywords: crop diversification, sustainable agriculture, nutritional security, food security

1. Introduction

An ever-increasing worldwide population, especially in many developing nations, necessitates additional food, fiber, and oil supplies, posing a serious challenge to agricultural scientists to produce more and more from limited, diminishing, and degraded land and water resources. By 2050, it is expected that the global population will have increased by 50%, and global grain demand would have doubled [1]. The stress from climate change, accompanying extreme weather and urbanization also creates the burden. Global agriculture in the present status points to a formidable challenge to agricultural sustainability. The most important danger to food security and the environment is dwindling per capita natural resources, as well as resource depletion and degradation. Existing intensification technologies are showing symptoms of wear and tear. The loss of biodiversity, groundwater shortages, fossil water extraction, groundwater contamination, and rising atmospheric CO_2 levels are all severe risks to sustainability. A variety of methodologies are used in sustainable production practises. Specific strategies must take into account the site specific and individual nature of sustainable agriculture. Reduced dependency on monocultures can give better resilience and reduce the chance of total system failure, which is critical for attaining long-term sustainable agricultural development. It can be a dynamic and continuous process to adjust in changing circumstances. Diversification is the process of utilization of the various emerging opportunities created by new market, technology, changes in governmental policies, higher profitability and also stability in the production system [2]. It is a useful strategy for reducing the risk in farming [3]. Crop diversification is generally viewed as shift from a traditionally grown less remunerative crops to more remunerative crops. Crop diversification is recognized as one of the most environmentally feasible, cost-effective, and reasonable approaches to reduce uncertainty in agriculture, particularly in the face of climate change. Crop diversification helps in minimizing the alleviating second generations problem such as soil degradation, soil salinity, insect-pest and disease insurgence, environmental pollution, decline in farm profit, nutrient imbalance, climate change etc. Crop diversification promotes farm resilience, or the ability of an agroecosystem to return to its former productive state after being perturbed, by increasing geographical and temporal biodiversity. Although crop diversification is not a new concept to many rural people in developing and emerging economies, there has been little research on the subject to date. However, there is increasing global interest in the area, owing to current worries about biodiversity loss, as well as human and environmental health. Thus, in this book chapter we are trying to give some understanding about the topic Crop diversification an effective strategy for sustainable agriculture development.

2. Concept of crop diversification

Crop diversification, as opposed to specialized farming, can be defined as an attempt to promote crop diversity by crop rotation, multiple cropping, or intercropping, with the goal of improving productivity, sustainability, and supply of ecological systems [4–6]. It could be one step toward more sustainable production systems, value chains for minor crops [7], and socioeconomic benefits [8]. Enhanced agricultural diversity, better diverse crop rotations, mixed cropping [9, 10], cultivation of grain legumes in generally cereal-dominated systems [11], perennial leys or grassland [12], and regionally adapted varieties or variety combinations are all examples of agricultural diversification strategies. In developing countries, crop diversification is defined as the substitution of one or more agricultural products for another. Diversification in agriculture can be defined as the reinvestment of some farm productive resources, such as land, capital, farm equipment, and labour, into new enterprises [13]. A shift from less profitable cropping system to more profitable cropping system is also known as diversification. Diversification of agriculture, in general, refers to transitioning from a single crop's regional or temporal dominance to the production of a variety of crops in order to meet the ever-increasing need for cereals, pulses, oilseeds, fibers, fuel, and feed. Crop diversification is a demand-driven, need-based situation specific and national goal seeking dynamic and iterative concept that incorporates

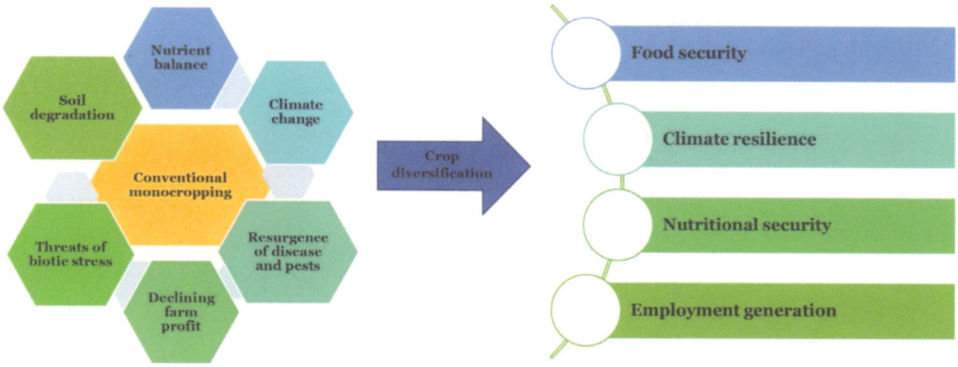

Figure 1.
Basic concept of crop diversification.

spatial, temporal, value addition, and resource-complementary techniques, as well as a move from traditional and less-remunerative crops (**Figure 1**).

3. Trends of agricultural diversification in South Asian Countries

South Asia has a long history of intensive agriculture, particularly irrigated rice cultivation techniques. Sector strategies in the region are mostly based on food self-sufficiency policies [14]. Throughout the last 30 years, the system's research and agricultural support services have increased food production faster than population expansion and diminished the percentage of people living in poverty. There has been significant income increase, diet diversification, and decreases in per capita grain intake throughout the comparable time span. South Asian countries are actively diversifying their economies in favor of high-value commodities such as fruits, vegetables, livestock, and fisheries, with some inter-country variation. Price policy, infrastructure development (particularly markets and highways), urbanization, and technical advancements all have a significant impact on agricultural diversification. Agricultural diversification in favor of high-value crops by substituting inferior coarse grains has helped rainfed areas more [15]. Agricultural diversification is also helping to increase export markets and create new job possibilities. Using appropriate institutions, it is necessary to properly coordinate the production and selling of high-value commodities. Market reforms in the form of building and strengthening desired institutions through necessary legal changes might go a long way toward encouraging agricultural growth, increasing small farm income, and boosting exports. Diversifying rural production is the process by which families create several livelihoods utilizing different variations of resources and assets in order to be less influenced by changes in the marketplace (such as price decreases) and to secure market stability [16]. So, if a region has high demographic pressure but minimal diversification, low-profit traditional commodities cultivation will increase and the farming frontier will spread, causing deforestation and soil erosion [17, 18]. As a result, investing in agricultural diversification can help to prevent environmental degradation by allowing for the production of a wider range of commercially feasible and productive crops [19]. Various options of crop diversification in South Asian countries are presented in the below **Figure 2**.

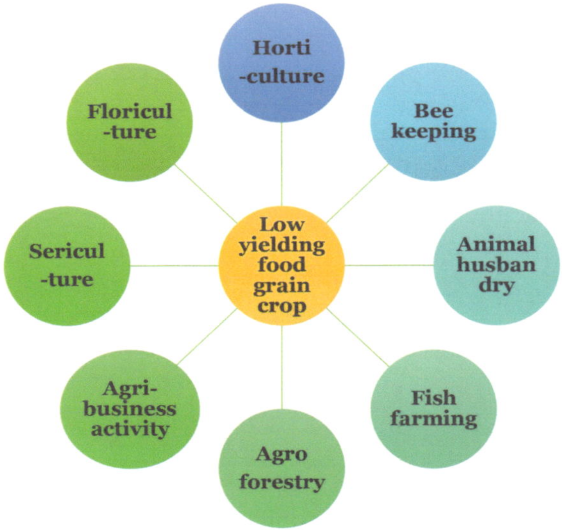

Figure 2.
Various options of crop diversification.

4. Approaches to crop diversification

The next sections examine the many techniques to crop diversification depending on land appropriateness, water availability, and market demand viz. regional, seasonal, and temporal [20]. The different approaches of crop diversifications are presented in **Figure 3**.

4.1 Horizontal diversification

It is done by basically two approaches, through crop substitution and crop intensification. These two approaches have been the two main process of crop diversification. Crop substitution means replacing any crop which is continuously growing as a monoculture crop or gain a tendency of specialization. For example, during green revolution era there was a tendency to growing cereals crops only. Now a days the trend has change a lot in developing countries. Farmers are shifting from monoculture cereals based staple food to high value crops like vegetable, spices etc. There are several advantages of crop substitution which could be higher net returns, improve resource use efficiency (land and labour), break in cycle of pest and disease etc. On

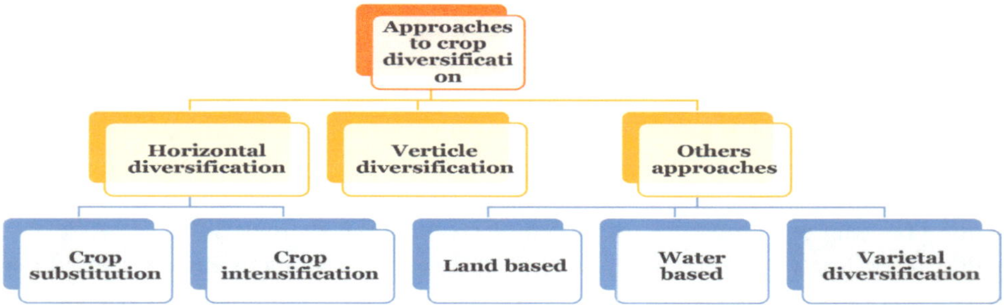

Figure 3.
Different approaches of crop diversifications.

Conventional cropping system	Crop intensification	Advantages	References
Maize-fallow	Maize–rajmash Maize–toria Maize–buckwheat Maize–buckwheat Maize (green cobs)-urdbean–buckwheat	Increased the grain equivalent yield, system production efficiency, relative production efficiency and land use efficiency.	Babu et al. [21]
Transplanted boro-transplanted aman	Wheat-mungbean-T. aman with full tillage Wheat-mungbean- dry seeded aman with strip tillage	Increased land and water productivity, system productivity.	Alam et al. [22]

Table 1.
Example of crop intensification and their advantages.

the other hand, crop intensification is adding of new value crops to existing cropping system to increase the farm's overall productivity. To reap the benefits of agricultural diversification, we must move away from simple crop rotation and toward intensive systems such as multiple cropping, intercropping, relay cropping, and so on. Crop intensification helps in job opportunity, profitability and energy use efficiency [21]. Some examples of crop intensification and their advantages are discussed in **Table 1**.

4.2 Vertical diversification

Vertical crop diversification, on the other hand, represents the degree and level of industrialization of agricultural production. In this approach famers and others add value to products through packaging, processing, regional branding, merchandizing to improve the marketable value of crops. Food crop vertical diversification is also described as the extension of post-harvest activities, such as processing and transformation industries, to allow food crops to be sorted, graded, processed into both food and industrial products, packed, stored, and transported to domestic or export markets [23]. The rise of processing and transformation industries appears to be the most important factor in rural areas in terms of creating revenue and jobs. To boost crop yields and income creation at the local, regional, and national levels, both types of diversification (*i.e.,* multiple cropping or horizontal diversification and agri-business or vertical diversification) will be required. The concept of vertical diversification is presented in the **Figure 4.**

4.3 Others approaches

- Land based approach
- Water-based approach
- Varietal diversification
- Diversification for nutritional security
- Diversification for nutrient management
- Diversification for pes management
- Diversification for mitigation and adaption of climate change

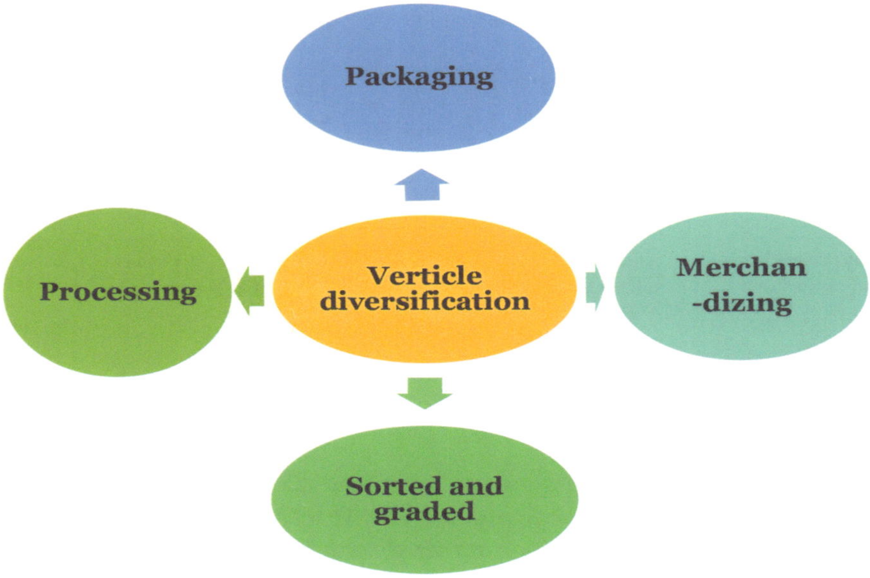

Figure 4.
Options of vertical diversification.

5. Measure of crop diversification and its characterization

Different measurements of crop diversification and their characterization are depicted in the **Table 2** [24].

Measure of crop diversification	Characterization
1. Temporal crop diversification	
Crop rotation	Growing of two or more different crops by one after another in consecutive ways
Catch crop	Growing of crops to in between the space of two main crop or when no main crops are being grown
Double or multiple cropping	Growing two or more crops in one growing season
Relay cropping	In relay cropping second crop is grown in standing crop before the first crop is harvested
2. Spatial crop diversification	
Alley cropping	It is an agroforestry system in which food crops are grown in alleys formed by trees
Intercropping	Growing two or more crops simultaneously on the same land with definite pattern
Mixed cropping	Growing two or more crops simultaneously in the same field
Variety mixture	Growing two or more varieties of a same species
Trap	Growing commercial and non-commercial crop simultaneously in the same land

Table 2.
Measure of crop diversification and its characterization.

Country	Sympson index of diversification in triennium ending			Sources of diversification (%) (1991–1992 to 1999–2001)	
	1981–1982	1991–1992	1999–2000	Cropping intensity	Crop substitution
Bangladesh	0.39	0.36	0.35	64.67	35.33
Bhutan	0.37	0.48	0.44	97.82	2.18
India	0.61	0.65	0.66	36.63	63.37
Maldives	0.77	0.77	0.77	83.22	16.78
Nepal	0.39	0.40	0.41	84.79	15.21
Pakistan	0.54	0.56	0.57	76.56	23.44
Sri Lanka	0.76	0.77	0.75	78.90	21.10
South Asia	0.59	0.63	0.64	42.98	57.02

Table 3.
Extent of diversification and sources of diversification in South Asian countries.

6. Crop diversification pattern in South Asian Countries

Extent of crop diversification pattern, Sympson index and sources of crop diversification is presented in **Table 3** [15].

7. Major driving forces for crop diversification

High-value commodity production is driven by demand, which is primarily determined by rising income and urbanization. The major drivers of crop diversifications are discussed in **Figure 5**.

1. Rapid urbanization of developing countries is one of the biggest reasons of crop diversification. Urbanization puts pressure on land resources, a small number of farmers requires to produce for a larger number of consumers.

2. Change in consumers demand due to shifting from a diet-based staple to nutrient rich animal products, fruits and vegetables.

3. Improving nutritional benefits by diversifying the monoculture of traditional cereals crop.

4. Climate change

5. Value addition

6. Export potential

7. The key driver in altering production portfolios in favor of high-value commodities is road and market. They connect the producer and the consumer directly, reducing transportation and transaction costs. Mostly in case of perishable items, they lessen the danger of post-harvest loss [15].

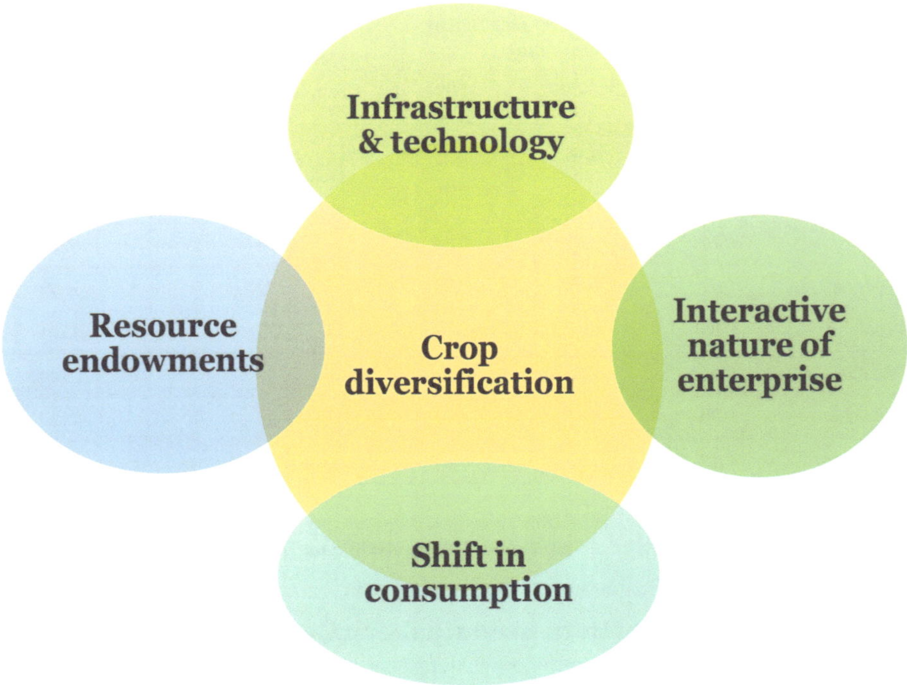

Figure 5.
Factors determining crop diversification.

8. Technology innovation may be a powerful driver for fostering agricultural diversification and accelerating agricultural growth. The fundamental driver of the 'Green Revolution' of the 1970s was biological technology [15].

9. Changing in governmental policy

10. Resilience and stability in production system.

11. Higher profitability

8. Need for crop diversification

1. Nutritional food security and quality of life can be improved through diversification in food basket.

2. Food security

3. Poverty alleviation

4. Employment generation

5. Trade needs

6. Protecting the environmental degradation by reversing the decline trend in soil productivity and ground water table.

7. Income growth

8. Ecological balance

9. Sustainability of natural resources

9. Strategies for crop diversification

1. Shifting from low yielding low value crops to high yielding high value crops.

2. Shifting toward higher water requirement crop to lower requirement crops.

3. Shifting toward low energy efficient crop to higher energy crop

4. Inclusion of legumes and oilseed crops

5. Inclusion of crop which has national and international market demand.

10. Advantages of crop diversification

10.1 Increasing the profitability of small farm holdings

The domination of marginal and small farmers is one of the primary issues confronting India's agricultural sector. These household makes up the majority of the rural population. Due to their low operating base, increasing the production of existing crops (staple food crops) may not be enough to boost their earnings. Therefore, diversifying the traditional cropping system is a best option to enhance income of small and marginal farmers.

10.2 Employment generation

Employment generation is a significant role of agriculture. But adopting the conventional cropping system like rice-wheat generally leads to lack of employment during off seasons. According to a number of studies, there is a serious problem of seasonal unemployment in different regions of our country, which leads to seasonal migration of labours/farmers to surrounding cities/towns in quest of contractual work [25]. Crop diversification helps rural households to have more opportunities of full-time employment.

10.3 Natural resource conservation and enhancement

Diversification is required to recover and enhance the value of the deteriorated natural resource base. Farmers in eastern India, particularly in West Bengal adopted wheat into a primarily rice system to take advantage of leftover moisture and so minimizes the need for wheat irrigation. In Punjab, on the other hand, an injudicious crop-mix, such as wheat-rice, has exacerbated the problem of water logging and salinity.

10.4 Improving export potential

To increase export potential, it is very much essential to adopt diversification in cropping systems. Such factors have weighed heavily on the minds of farmers in eastern India, particularly in West Bengal, where wheat has been introduced into a primarily rice system to take advantage of leftover moisture and so minimizes the need for wheat irrigation.

10.5 Risk reduction

Crop diversification is very much responsive to climatic and biotic vagaries, particularly in fragile ecosystems by expanding locally adapted or introducing novel varieties and related production systems will help resource-poor farmers improve their food security and income generation while also protecting the environment [26].

10.6 Pest and disease control

Crop diversification, which favors species combinations over monocultures, is one of the most cost-effective ways to combat pests and disease, and it has sparked a lot of attention in recent years [27].

10.7 Improvement of soil fertility

One of the most important constraints for sustainable crop production is low soil fertility. In smallholder systems, poor farming practises, mostly continuous cropping with limited external inputs, have gradually depleted soil fertility. Interaction of crop species with beneficial soil biota helps in maintaining biogeochemical cycling of both organic and inorganic nutrients in the soil and maintaining soil quality [28].

11. Review of literature

Kasem and Thapa during 2011 conducted a study in Thailand, collecting primary data from 245 farm households using a structured questionnaire to examine the impact of crop diversification on income and input consumption. They discovered that the vast majority of farmers stated that crop diversification contributed to a significant rise in their revenue [29]. The results of their research findings are depicted in **Table 4.**

Birthal et al. studied into the impact of crop diversification on India's farm poverty. Data from a nationally representative survey was used. The dataset, according to them, contains information on the crops grown, as well as the costs and returns

Opinion	Frequency (n = 81)	%
Increased income	68	84
Enhanced food sufficiency	54	66.7
Flow of income throughout the year	43	53.1
Offers opportunity to produce crops according to market demand	12	14.8
Smoothens the effect of price fluctuation	10	12.3

Table 4.
Diversified farmers viewpoint about benefits of crop diversification.

Crops	Marginal ≤1 ha	Small (1–2 ha)	Medium (2–4 ha)	Large >4 ha	All
Total cereal	9044 (456)	7099 (256)	7518 (403)	6164 (599)	8301 (304)
Fruits	37,347 (9283)	51,859 (19,187)	36,726 (13,289)	30,433 (13,585)	39,523 (9566)
Vegetable	22,423 (3100)	19,226 (1748)	20,641 (2402)	19,114 (4657)	21,459 (1852)
High value crops	25,618 (2486)	22,329 (2292)	21,411 (2834)	21,518 (4014)	24,263 (2091)

*Figures in parentheses are standard errors. Total cereals include rice, wheat, maize, and coarse cereals like pearl millet, sorghum, and barley. High-value crops include vegetables, fruits, condiments and spices, flowers, aromatic and medicinal plants, and plantation crops like tea and coffee.*One US$ = 47.62 in the survey year i.e., 2002–2003 [30].*

Table 5.
Comparison of net returns (Rs ha^{-1}) from higher value crops with other crops by crop diversification.

associated with each crop. This allows us to investigate the pattern and breadth of high value crop diversification across land sizes, as well as their profitability in comparison to other crops. In comparison to other crops, **Table 5** shows the estimated net returns per hectare from high value crop cultivation. When compared to cereals, high value crop (HCVs) provided much higher returns to all types of farmers, including marginal farmers [30].

Despite differences between countries, rural households in the majority of countries tend to rotate a small number of crops. Two, three, or a maximum of four agricultural products are the most common combinations used by households. Few households grow more than six distinct crops, most likely due to the small size of their allotment and the inherent challenge of producing many goods viz. water requirements, necessity of sun exposition and type of soil, among others. An empirical evidenced from eight different countries were analyzed and presented in **Table 6** [31].

Diversification of crop through intercropping system has significant advantage in land use efficiency, monetary returns and crop productivity as compared to monocropping. Intercropping results in more efficient use of solar energy and harnessing benefits of positive interactions of crop association. Benefits of some potential intercropping system are discussed in below **Table 7** with regards to system productivity, net returns and B:C ratio.

Country and year	Number of crops produced and share of households (% of total national sample) producing each number								
	1	2	3	4	5	6	7	≥8	Total
Malawi, 2004	11	21	23	20	13	6	3	3	100
Nepal, 2003	3	25	8	18	8	10	3	25	100
Vietnam, 1998	7	7	8	8	9	7	8	46	100
Pakistan, 2001	22	61	15	2	0	0	0	0	100
Nicaragua, 2001	6	19	20	17	11	9	7	11	100
Indonesia, 2000	28	29	25	11	4	2	1	0	100
Albania, 2005	11	31	15	14	8	9	3	9	100
Panama, 2003	36	38	19	6	1	0	0	0	100

Table 6.
Share of household practicing different numbers of crops (an empirical evidence from eight developing countries) [31].

Intercropping system	Location	System productivity (t ha^{-1})	Net returns (×10^3 ₹ ha^{-1})	B:C ratio	References
Chickpea + Indian mustard	Kanpur, India	2.4	17.1	2.4	[32]
Sugarcane + Maize	Pantnagar, India	200.6	124.9	1.90	[33]
Wheat + Mustard	Kangra, India	4.7	26.7	2.55	[34]
Maize + Potato	Pusa, New Delhi	14.0	35.7	2.14	[35]
Ratoon cane + Berseem	Lucknow	90.8	56.2	2.64	[36]

Table 7.
Economics of intercropping system for crop diversification.

12. Constrains of crop diversification

These are primarily socioeconomic and institutional barriers, such as the lack of holding consolidation and group farming, geographic disadvantages (remote areas far from shops and supermarkets), farmer 'lack of education, the outright failure of the agricultural extension system, and a lack of transportation and marketing facilities.

1. Lack of salt and excess moisture tolerant crops and cultivars.

2. Lack of skill and knowledge in choosing alternate crops in cropping system

3. Small and fragmented land holding creates difficulty to ensure that they participate more fully in crop diversification.

4. Agricultural output is used as a raw material in agro-based industries. When monoculture becomes unsustainable, a more sustainable and profitable crop must be substituted. Because of massive infrastructure expenditure, switching over becomes difficult by that time; for example, the rice industry in Punjab and Haryana, the sugarcane industry in Uttar Pradesh, and the soybean industry in Madhya Pradesh states in India.

5. The major causes of high cost of production are rising wage rates and declining factor productivity. The researchers are being challenged to reduce the cost of production and produce new adaptive cultivars that can capture high market prices.

6. Over use and sub optimal use of natural resources like water and land resources, may negative impact on environment and sustainability.

7. Weak research-extention and farmers linkage.

8. Lack of knowledge among the farmer

13. Lack of concept of Crop diversification

Though there are hundreds of scientific papers in the field of agronomy on agricultural diversity such as crop rotation or intercropping, only a small percentage of these studies are about diversification as a concept [21].

14. Conclusion

Diversification is one of the most effective ways to boost farm revenue, resulting in increased food, nutrition, and environmental security, as well as poverty reduction in developing countries. It creates a tremendous impact on agro-socio-economic gains.

- It increased the flow of income throughout the year.
- Offers opportunity to produce crops according to market demand
- Smoothens the effect of price fluctuation
- Increase the grain equivalent yield, system production efficiency, relative production efficiency and land use efficiency of maize-fallow system.

15. Future prospects for the adoption of crop diversification

- Overall potential of crop diversification is yet to be studied.
- Impact of crop diversification on rural economics and poverty alleviation needs to be investigated in details.
- Effect of crop diversification on soil health properties needs to be studied in details.
- Social benefits of crop diversification are less well known.

Author details

Anamika Barman[1*], Priyanka Saha[1], Shashank Patel[1] and Anurag Bera[2]

1 Division of Agronomy, ICAR-Indian Agricultural Research Institute, New Delhi, India

2 Department of Agronomy, Dr. Rajendra Prasad Central Agricultural University, Pusa, Bihar, India

*Address all correspondence to: anamikaiari123@gmail.com

References

[1] Godfray HC, Beddington JR, Crute IR, Haddad L, Lawrence D, Muir JF, et al. Food security: The challenge of feeding 9 billion people. Science. 2010;**327**(5967):812-818

[2] Chand R, Chauhan S. Socio economic factors in agricultural diversification in India. Agricultural Situation in India. 2002;**58**(11):523-530

[3] Hedge DM, Tiwari SP, Rai M. Crop diversification in Indian agriculture. Agricultural Situation in India. 2003;**60**:255-272

[4] Kremen C, Miles AF. Comparing biologically diversified with conventional farming systems: What is known about environmental benefits, externalities and tradeoffs among crop productivity and ecosystem services. Ecology and Society. 2012;**17**:40

[5] Garbach K, Milder JC, DeClerck FA, Montenegro de Wit M, Driscoll L, Gemmill-Herren B. Examining multi-functionality for crop yield and ecosystem services in five systems of agroecological intensification. International Journal of Agricultural Sustainability. 2017;**15**(1):11-28

[6] Wezel A, Casagrande M, Celette F, Vian JF, Ferrer A, Peigné J. Agroecological practices for sustainable agriculture. A review. Agronomy for Sustainable Development. 2014;**34**(1):1-20

[7] Meynard JM, Charrier F, Le Bail M, Magrini MB, Charlier A, Messéan A. Socio-technical lock-in hinders crop diversification in France. Agronomy for Sustainable Development. 2018;**38**(5):1-3

[8] Feliciano D. A review on the contribution of crop diversification to sustainable development goal 1 "No poverty" in different world regions. Sustainable Development. 2019;**27**(4):795-808

[9] Bedoussac L, Journet EP, Hauggaard-Nielsen H, Naudin C, Corre-Hellou G, Jensen ES, et al. Ecological principles underlying the increase of productivity achieved by cereal-grain legume intercrops in organic farming. A review. Agronomy for Sustainable Development. 2015;**35**(3):911-935

[10] Malézieux E, Crozat Y, Dupraz C, Laurans M, Makowski D, Ozier-Lafontaine H, et al. Mixing plant species in cropping systems: Concepts, tools and models: A review. In: Sustainable Agriculture. Springer; 2009. pp. 329-353

[11] Watson CA, Reckling M, Preissel S, Bachinger J, Bergkvist G, Kuhlman T, et al. Grain legume production and use in European agricultural systems. Advances in Agronomy. 2017;**144**:235-303

[12] Haughey E, Suter M, Hofer D, Hoekstra NJ, McElwain JC, Lüscher A, et al. Higher species richness enhances yield stability in intensively managed grasslands with experimental disturbance. Scientific Reports. 2018;**8**(1):1-10

[13] Gangwar B, Duhoon SK, Gill MS. System based production potential, yield gaps and diversification opportunities in selected zones of India. Progressive Agriculture. 2010;**10**(3):19-26

[14] Dorjee K, Broca S, Pingali P. Diversification in South Asian agriculture: Trends and constraints in South Asia: Patterns, determinants

and policy implications. Economic and political weekly. In: Agricultural Diversification and Smallholders in South Asia. New Delhi, India: Academic Foundation Publishers; 2007. pp. 129-150

[15] Joshi PK, Gulati A, Birthal PS, Tewari L. Agriculture diversification in South Asia: Patterns, determinants and policy implications. Economic and Political Weekly. 2004;**39**:2457-2467

[16] Niehof A. The significance of diversification for rural livelihood systems. Food Policy. 2004;**29**(4):321-338

[17] Angel A, Ramos H. Competitividad de alternativas para la diversificación Agrícola: Frutas y Hortalizas. In: Proyecto CRECER/USAID. San Salvador: Chemonics International, Inc.; 1997

[18] Barrett CB, Bulte EH, Ferraro P, Wunder S. Economic instruments for nature conservation. Key Topics in Conservation Biology. 2013;**2**(2013):59-73

[19] Di Falco S, Perrings C. Crop biodiversity, risk management and the implications of agricultural assistance. Ecological Economics. 2005;**55**(4):459-466

[20] Nishan MA. Crop diversification for sustainability. Popular Kheti. 2014;**2**(2):20-24

[21] Babu S, Singh R, Avasthe RK, Yadav GS, Rajkhowa DJ. Intensification of maize (Zea mays)–based cropping sequence in rainfed ecosystem of Sikkim Himalayas for improving system productivity, profitability, employment generation and energy-use efficiency under organic management condition. Indian Journal of Agricultural Sciences. 2016;**86**(6):778-784

[22] Alam MJ, Humphreys E, Sarkar MA. Intensification and diversification increase land and water productivity and profitability of rice-based cropping systems on the high Ganges River Floodplain of Bangladesh. Field Crops Research. 2017;**209**:10-26

[23] Hedley DD. Diversification: Concepts and directions in Indonesian agricultural policy. Soybean Research and Development in Indonesia. 1987;**10**:17

[24] Hufnagel J, Reckling M, Ewert F. Diverse approaches to crop diversification in agricultural research. A review. Agronomy for Sustainable Development. 2020;**40**(2):1-7

[25] Behera UK, Jha KP. Agro-economic assessment of different crops and cropping systems for the drought prone uplands of Kalahandi, Orissa. Journal of Andaman Science Association. 1999;**15**(1):30-34

[26] Shoffner AV, Tooker JF. The potential of genotypically diverse cultivar mixtures to moderate aphid populations in wheat (*Triticum aestivum* L.). arthropod-plant. Interactions. 2013;**7**(1):33-43

[27] Tooker JF, Frank SD. Genotypically diverse cultivar mixtures for insect pest management and increased crop yields. Journal of Applied Ecology. 2012;**49**(5):974-985

[28] Jeffries P, Gianinazzi S, Perotto S, Turnau K, Barea JM. The contribution of arbuscular mycorrhizal fungi in sustainable maintenance of plant health and soil fertility. Biology and Fertility of Soils. 2003;**37**(1):1-6

[29] Kasem S, Thapa GB. Crop diversification in Thailand: Status, determinants, and effects on income and use of inputs. Land Use Policy. 2011;**28**(3):618-628

[30] Birthal PS, Roy D, Negi DS. Assessing the impact of crop diversification

on farm poverty in India. World Development. 2015;**72**:70-92

[31] Pellegrini L, Tasciotti L. Crop diversification, dietary diversity and agricultural income: Empirical evidence from eight developing countries. Canadian Journal of Development Studies/Revue canadienne d'études du développement. 2014;**35**(2):211-227

[32] Tripathi HN, Chand S, Tripathi AK. Biological and economical feasibility of chickpea (*Cicer arietinum*) + Indian mustard (*Brassica juncea*) cropping systems under varying levels of phosphorus. Indian Journal of Agronomy. 2005;**50**(1):31-34

[33] Rana NS, Kumar S, Saini SK, Panwar GS. Production potential and profitability of autumn sugarcane-based intercropping systems as influenced by intercrops and row spacing. Indian Journal of Agronomy. 2006;**51**(1):31-33

[34] Kumar A, Thakur KS. Effect of Brassica spp. and their sowing proportions on productivity, competition and economics of wheat (*Triticum aestivum*) + Brassica mixed cropping system under rainfed conditions of Himachal Pradesh. Indian Journal of Agronomy. 2006;**51**(4):259-262

[35] Bharati V, Nandan R, Kumar V, Pandey IB. Effect of irrigation levels on yield, water-use efficiency and economics of winter maize (*Zea mays*)-based intercropping systems. Indian Journal of Agronomy. 2007;**52**(1):27-30

[36] Singh AK, Lal M, Prasad SR, Srivastava TK. Productivity and profitability of winter-initiated sugarcane (Saccharum spp. hybrid complex) ratoon through intercropping of forage legumes and nitrogen nutrition. Indian Journal of Agronomy. 2007;**52**(3):208-211

Section 3

Soil Management

Chapter 6

Reducing Soil Compaction from Equipment to Enhance Agricultural Sustainability

Michael M. Boland, Young U. Choi, Daniel G. Foley, Matthew S. Gobel, Nathan C. Sprague, Santiago Guevara-Ocana, Yury A. Kuleshov and Robert M. Stwalley III

Abstract

The compaction of agricultural soils cannot be solved, only managed. As a compressible media, soil travel without causing some collapse of the existing structure is impossible. If left uncorrected, farmers can see up to a 50% reduction in yield from long-term compaction. This chapter will describe the effects of soil compaction on the environment, crop quality, and economic sustainability. The base causes will be examined, along with the engineering designs for vehicles that minimize the problem. The tracks versus tires debate will be thoroughly discussed, and the advantages and disadvantages of each system will be detailed. It will be shown that although tires represent the likely current best economic option for vehicle support, the potential of tracks to reduce compaction has been fully exploited. The advantages of four-wheel drive vehicles in reducing soil compaction will be shown, along with the mitigation potential of independently driven wheels and active soil interaction feedback loops. The design of crop production tillage equipment and tillage tool working points will be explored, along with the concept of critical tillage depth. Equipment for compaction relief will also be discussed, as will the sustainable agricultural protocols of cover crops, crop rotation, and controlled traffic farming.

Keywords: agricultural tillage, compaction remediation, cover crops, off-road vehicle design, tires, tracks, tractors, soil compaction, sustainability

1. Introduction

Since the late 1960s, the agricultural industry has taken an increasing interest in the effects of soil compaction on soil health, agricultural practices, water runoff, and the sustainability of grain production. Compaction results from any practice that includes traveling over the soil. This can be caused by heavy-axle machinery, excessive ground working, livestock, or specific geotechnical practices, such as rolling, which is used to compact the soil in preparation for construction. Repeated soil compaction experiences have cumulative negative effects for agricultural soils, such as a decrease in pore space, reduced pore nutrient and water uptake, denitrification, and enhanced difficulties in seed germination. The effects of compaction also extend beyond agriculture and are of concern to environmental specialists all over the world. For instance, high compaction rates increase the likelihood of water retention issues, water runoff, and erosion. From the last 60 years of research, modern agricultural operations have progressed to incorporate a variety of soil compaction reducing approaches. These approaches include equipment solutions like tracked implements, happy-seeders, and complex multi-crop planters that reduce field traffic. Within the scope of production agriculture, many existing practices unrelated to vehicle design, like no-till seeding, have decreased the impact of soil compaction and help to repair damaged and heavily compacted soils. These design improvements and management practices will be explored in this chapter, and their effectiveness will be measured. This topic is particularly timely and relevant because present-day tractors have increased in size compared to traditional row crop tractors for better productivity and field efficiency. Although most smaller-scale agricultural equipment is used for multiple tasks, the presence of a variety of different off-road vehicles on the market indicates a broad need for various equipment types and provides an opportunity for exploring the existing and potential solutions to soil compaction problems in different off-road vehicle designs. The chapter will proceed with an analysis of how soil compaction is addressed in machine design, as well as new areas that deserve more specific research and improvement. Multiple factors are involved in soil compaction, and multiple designs exist to address these various factors. The present off-road vehicle offerings are clearly less than ideal for long-term soil health. There is a potential for improvement in existing designs to benefit all involved stakeholders, and this potential will be explored.

2. The relevance of soil compaction and its effects on sustainability

Interest in soil compaction dates back to the time when humans started to use draft animals as a main source of power in agriculture. Many authors have addressed the problem since the 19th century. One of the first recognitions of the problem in academic literature dates back to 1857, with a description of the Fowler steam engine-powered plowing system [1]. Draft animals are still being used on vast areas of land in developing countries, and the animal-induced compaction problem continues to this day. The growing use of steam-powered tractors added to soil compaction concerns in the second half of the 19th century. While the mass-power ratio allowed for the widespread use of powerful tractors, the vehicles were still very heavy, and the need to minimize wheel impact on the soil was quickly recognized.

Different engineers have attempted to address the problem in multiple ways. These attempts did not lead to a unified design, but they moved the engineering thinking forward and were instrumental in creating the more successful designs of the 20th century. Between the last decade of the 19th century and 1904, internal combustion engines (ICE) replaced steam engines on tractors in America, and a new era of agriculture began. The better mass-power ratio of the ICE provided for lighter designs and less impact on the soil, but other problems emerged. Mass agriculture meant the more intensive use of fields and more impact on the soil per given period in time. One of the first experiments describing soil compaction problems was run in 1944. The topic has continued to be a strong focus for agricultural researchers. Raney, Edminster, and Allaway conducted the first review of literature on compaction in agricultural soils in America, which included 43 references [2]. The so-called "load index" started to grow at about the same time and exceeded that of the early 20th century by the 1970s. The soil compaction problem continues to be of a major focus of agricultural, industrial, and academic practitioners and researchers. One industrial example is Caterpillar's efforts to use tracks in agriculture in agriculture to decrease soil compaction [3]. Other modern solutions have attempted to address the problem, and the current review will introduce them to the reader [4–6].

2.1 Environmental impact of soil compaction

Soil compaction has a measurable influence on the environment, specifically on atmospheric, water, and soil resources. Agricultural operations have a major impact on the atmosphere through the emission of greenhouse gases. Soil "compaction may change the fluxes of these gases from the soil to the atmosphere because of its influence on soil permeability, soil aeration and crop development" [4, p. 8]. Water resources include both surface and ground water volumes. Soil compaction negatively affects the infiltration of different substances into the ground. Ammonia injected into the soil can escape into the atmosphere faster in a compacted soil than in an uncompacted one. Soil compaction also perpetuates the accumulation of rainwater on the surface in low parts of the field and increases the likelihood of runoff events. The latter leads to excessive sediment and chemical transfer into surface ground water resources, such as local rivers, lakes, ponds, and bigger regional natural water reservoirs [5, 7].

Soil biota is responsible for the decomposition of organic matter, release of nutrients and formation of aggregates [8]. Such tasks are performed by microfauna (bacteria, fungi), which are fed upon by meso- and macrofauna (protozoa, nematodes, arthropods) within the soil food web. **Figure 1** illustrates some of these various interconnections between the living things in the soil.

Soil compaction creates a rearrangement of soil particles that leads to a reduction of void space, a phenomenon that can be measured in several different ways. At first glance, there are visual and tactile methods that can provide a quick assessment, but to quantify the effects of soil compaction, physical parameters must be measured. Direct and indirect measures are used together to enable a deeper understanding of the characteristics of the total volume of the soil, such as bulk density (direct), soil strength (indirect), soil electrical resistivity (indirect) and water infiltration rate (indirect). **Figure 2** shows two examples of soil profiles exhibiting compaction effects. The soil on the left has a better structure above and below the compacted layer located between 10 *cm* and 40 *cm* depth [9]. On the right, a compacted layer in wetland creates a toxic environment for roots and soil biota. Soil compaction effects vary by location based on multiple interconnected factors, making a comprehensive assessment of specific fields the key to securing the sustainability of any agricultural operation over time.

2.2 Effects on harvest quality and farmlands

Soil compaction has a directly visible effect on the crop that is being grown in the degraded area. As soil compacts, it reaches a point of root growth restriction that is

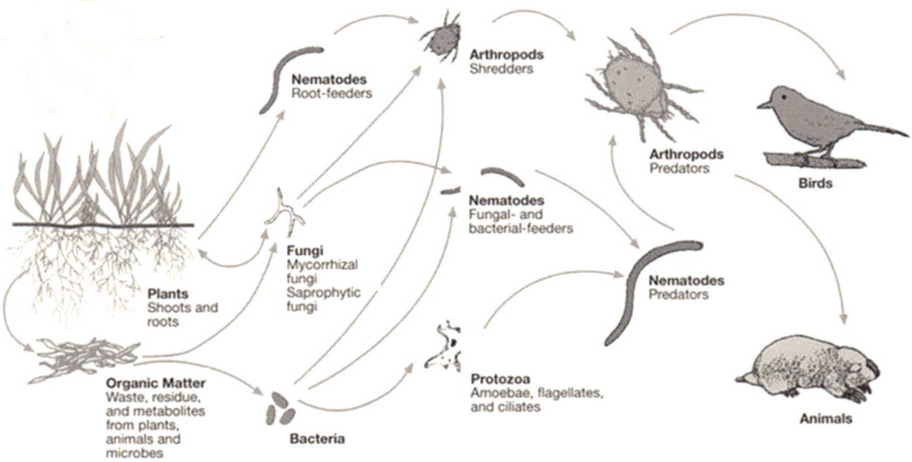

Figure 1.
Soil biota species and food web (duiker, 2005) – [8].

Figure 2.
A compacted layer under dryland canola (left), and a gray anaerobic layer in a clay loam soil (right) (Nawaz et al., 2013) – [9].

highly detrimental to both the quality and health of plants, as well as the quantity of the cultivated crop yield [10]. The lack of loose soil aggregates prevents strong root formations. This leaves crops more susceptible to wind and water damage. There is reduced nutrient uptake, since the root mass of the plant is diminished in both absorption volume, as well as effectiveness. Individual plants are less healthy and produce significantly less grain and forage mass. Perennial crops, like many fruit plants, stop root growth when confronted with significant compaction. Beyond the lack of void space, nutrient uptake in compacted topsoil is greatly reduced as the biological health of the soil diminishes [11]. Crops growing in densified soils can be expected to be brittle, due to the reduced nutrient intake as soil compaction causes reduced aerobic microbial activity and denitrification [12]. Soil compaction has a progressively negative effect on the biosphere. As the soil is compacted and continuously depleted, natural vegetation, such as weeds and grasses, quickly gets restricted from lack of soil aeration. The crushing of the soil and diminishing amount of additional biomass that would otherwise be introduced into the soil can eventually lead to an elimination of plant life causing an open and exposed soil surface. Soil in this condition is more easily impacted by wind and water erosion. If preventative measures are not taken, the effects of soil compaction on crop quality and farmland are cumulative, can take place quickly, and have lasting damage [13].

As shown by a review of the soil compaction literature [14], studies detailing the continuing long-term effect of compaction on a specific piece of ground are rare. However, as shown in **Figure 3**, it can take years for soil to naturally recover following a single compaction event [15]. Studies in cotton, as displayed in **Figure 4**, show a significant decline in crop yield within the initial season of the compaction event [16]. Since the effects of compaction are cumulative and continue from one season on into the next, it can be inferred that the decline from unmitigated soil compaction will continue to grow and magnify under the same management processes. **Figure 5** presents the general effect over time on production costs and gross margin of the farming operation [14].

2.3 Social and economic impact

Soil compaction has a negative impact on the economy of agricultural operations in the long-term. Soil compaction decreases the quantity and quality of harvest. Continuous and unaddressed soil compaction does not allow soil to sustainably recover through natural means [17, 18]. This affects the local food security in the regions where the traditional economy relies on agriculture. The local quality of life and general economic health of an agricultural region is adversely affected when local soils become compacted. Multiple potential solutions can help minimize the impact. Some come from farmer experience and depend on the operator in the field. Others are industry-wide, general practices. Academic researchers model the problem by studying economic impact. These models provide for better forecasting, equipment selection, and targeted problem solving. One model suggests that in the short-term, the negative impacts of soil compaction can be compensated by "more timely field

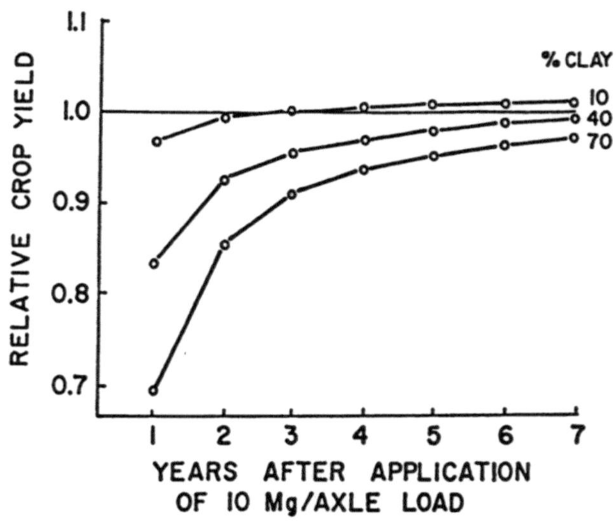

Figure 3.
Yield recovery following a significant compaction event (Voorhees, 1986) – [15].

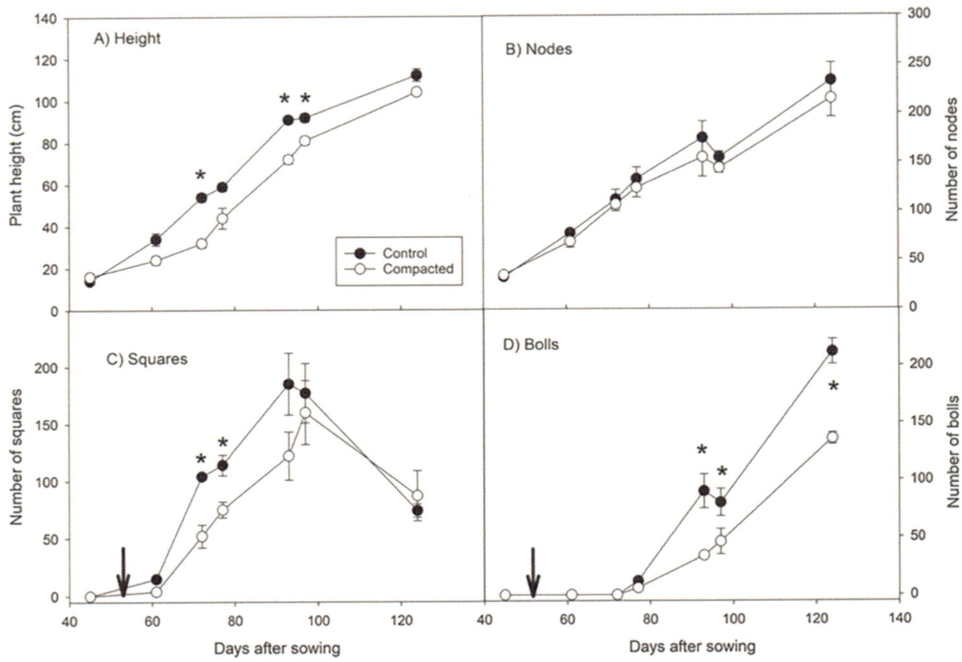

Figure 4.
Difference in same year cotton yield between compacted and uncompacted ground (Jamail et al., 2021) – [16].

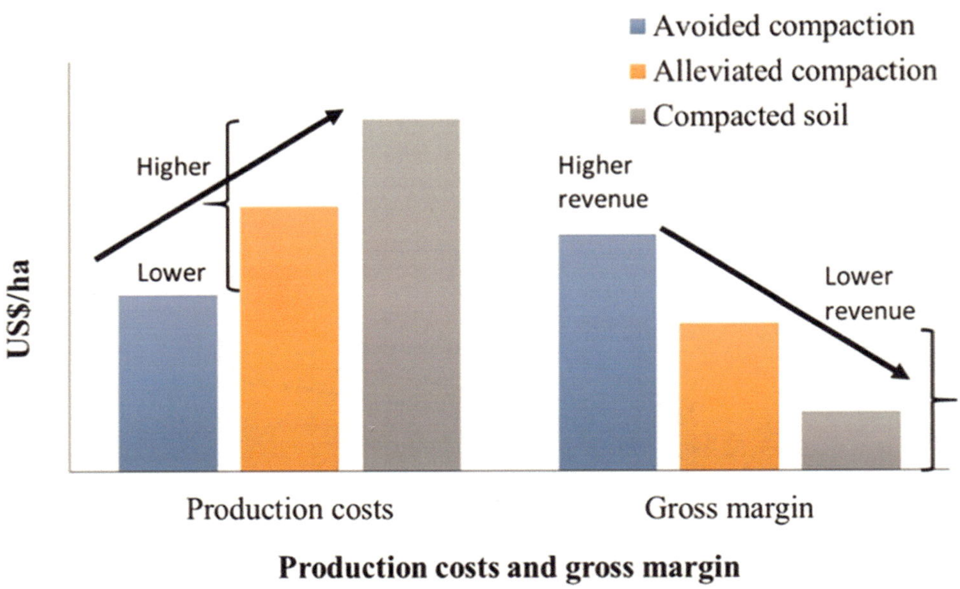

Figure 5.
The generalized trends for production cost and gross margin for avoided compaction, relieved compaction, and compacted soils in production (Chamen, 2015) – [17].

operations." Profits and productivity generate energy costs, air pollution, capital costs, timeliness costs, and soil erosion, which are also evaluated [19]. Other studies address the problem through even more sophisticated modeling. Additional effort is needed to standardize the corporate and academic researchers' efforts to improve these models for specific predictive tasks. Soil compaction models have the potential to help the businesses and governments to develop advanced solutions for real-world agricultural problems through an improved understanding of the social and economic impacts.

The extent of the economic effect of soil compaction is difficult to quantify as it is ultimately very situational. Under circumstances where soil requires additional operation to alleviate the effects of long-term compaction, the cost of crop production rapidly increases to unviability. Soil health does not always deteriorate to the point where intervention is required, but this does not mean that these farming operations are unaffected. The most common issue caused by soil compaction is the decrease in crop productivity. **Figure 6** below summarizes the impact soil compaction has on soil and crop health [20]. Reduction in plant growth and development, such as biomass accumulation, stomatal photosynthesis, and poor proliferation, as well as poor nutrient and water uptake decrease yield and overall crop productivity. **Figure 7** depicts the impact on potato yield resulting from varying irrigation levels [21]. This graph shows the availability of adequate water can increase yield by at least 100%. Because soil compaction so negatively impacts water availability and uptake, the conclusion can be drawn that compaction issues can decrease crop yield and productivity by up to 50%. In short, this also means that farmers risk losing 50% of expected profits, when the soil compaction problems are not properly addressed. Inattention to this vital issue in land management can destroy the resource's ability to be productive both now and in the future. It is imperative that farm managers understand the connection between management of soil compaction today and the long-term sustainability of the agricultural ground into the future.

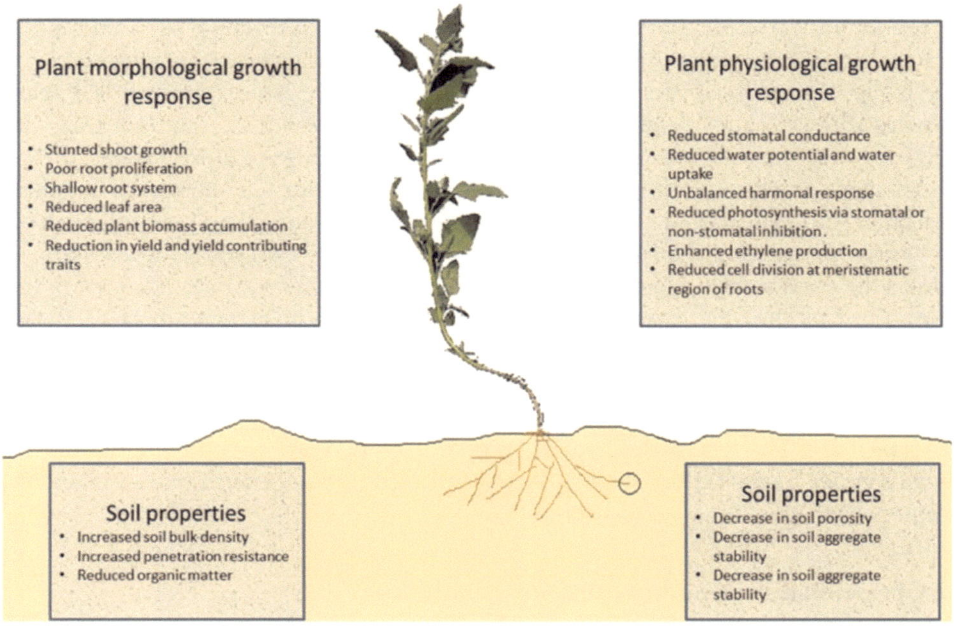

Figure 6.
Summary of the knowledge of the effects of soil compaction on soil plant morphological and physiological growth and soil properties (Shah et al., 2017) – [20].

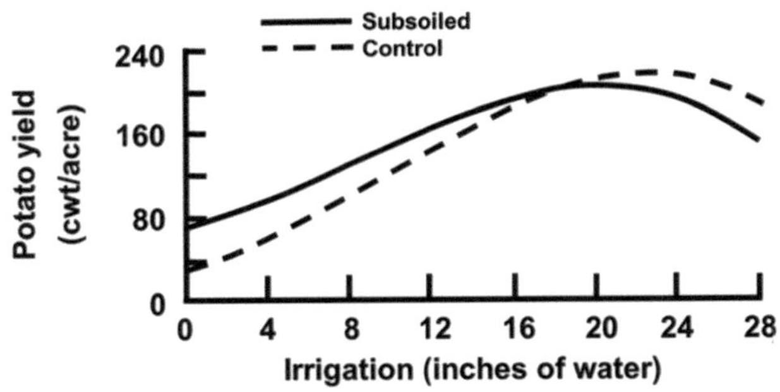

Figure 7.
Potato yield at different irrigation levels for subsoil and control fields (Ghosh & Daigh, 2020) – [21].

3. The causes of soil compaction

Soil compaction is the phenomenon associated with the collapse of soil media to support the loads imposed upon it. All agricultural operations on the surface of the ground cause soil compaction. Heavy axle loads, wet soil operations, livestock grazing, and materials stored directly on the surface can all result in unwanted compaction. The details of these agricultural process root causes of soil compaction will be explored in this section.

3.1 Operation of equipment with heavy axle loads

An axle load is the total load supported by a single axle, usually across two points of contact on either side of the vehicle. Although most agricultural equipment uses two axles for load distribution, each point of contact carries harmful loads into the soil. A large agricultural vehicle weighing 20 *ton*, creates 10 *ton* of force on each axle and causes the soil beneath each point to compact, until it can support the imposed load. The biggest factor to consider in reducing soil compaction is large axle loads. For two vehicles with the same weight distribution, the bigger the vehicle's contact area with soil, the lesser the pressure is applied to the soil surface. **Figure 28** illustrates an advantage of tracks over tires by the contact area parameter [22]. Research has shown that having an axle load of 10 *ton* can cause deep (more than 45 *cm*) subsoil compaction under moist conditions [8]. Grain carts and other heavy trailing implements behind the power units add to the problem of soil compaction, since axle load is determined by the total weight of the vehicle divided by the number of axles. Reducing single axle loads below five *ton* or less will diminish subsoil compaction, and only cause topsoil compaction [8]. Using heavy machinery under wet or moist conditions always increases soil compaction dramatically over use under dry conditions for most soil types [23]. The relationship among pressure applied, water content and bulk density varies across different soil types as particles rearrange with changing water contents [24].

3.2 Operation during non-optimal soil conditions

Under non-optimal soil conditions, field farm operations should be considered with great reluctance, due to the potential for severe damage to the soil matrix. As farm equipment crosses through a wet field, ruts are formed from soil compaction around the tire path. Tillage is a common practice to relieve soil compaction due to poor soil management. However, tilling breaks-apart the soil structure and causes further traffic, in addition to deeper compaction in the field. A tilled soil is more easily compacted, since the subsoil beneath the tillage line is now in a more vulnerable state for soil compaction [25]. Under good soil conditions, the integrity of the soil is reasonably strong and minimizes the loss of pore space from heavy equipment travel. When soil conditions are non-optimal, the structural integrity of the soil is significantly reduced, and this results in the elimination of pore space with vehicle traffic. As shown in **Figure 29**, when the same pressure is applied in a loam soil, the bulk density significantly increases with increasing soil water content, thus, leaving the soil susceptible to compaction [24]. Additionally, water within the soil matrix reduces the coefficient of friction between neighboring soil particles and promotes the ease of displacement and flowability of the soil.

3.3 Livestock grazing

Livestock grazing can affect soil stability and functionality if not managed properly. The severity of soil damage due to livestock grazing is related to the soil type, texture, and moisture content. Pugging, the formation of soil around the hoof of the livestock, can result in increased soil compaction and a reduction in soil surface water infiltration rates [26]. When water does not infiltrate through the soil surface during rainfall or irrigation, puddling occurs in fields. The trampling and pugging from livestock onto soil surfaces damages the subsurface soil integrity. The density of the livestock per unit of area in a pasture impacts the level of soil compaction due to pugging. This effect also negates the value of winter grazing on crop land to glean harvest losses. The long-term damage from soil compaction to the crop ground greatly outweighs the value of the "free" feed gained.

3.4 Other

Aside from intensive farming and grazing practices common in modern agriculture, there are other factors, some environmental and some man-made, that can have a noticeable effect on soil compaction. Depending on the region of agricultural production, the type of soils, as well as natural and artificial drainage, some fields can be subject to prolonged ponding of water in localized areas. Over time, the weight of the water ponded on the soil surface causes the soil pores to collapse further, slowing the movement of water through the soil and increasing the weight of water on top of the soil surface during future precipitation events. Water ponded on the soil surface adds 10 *kPa* of pressure per m of depth. Additionally, slowed water movement through the soil increases the risk of farming operations occurring during non-optimal soil conditions. Another non-conventional contribution to soil compaction is the relatively new practice of storing grain in large plastic bags that are laid-out on the soil surface. Producers using this method of temporary grain storage have noted significant soil compaction on the surface due to the weight of the grain.

4. Off-road vehicle designs for soil compaction management

Agricultural tractive power units are the largest source of unwanted soil compaction today. Significant research and financial investment have been made in methodologies to reduce the compaction from these vehicles. Tracked systems and advanced tire systems are both designed to spread the loads imposed upon the soil below detrimental levels. This section will review these common undercarriage systems, along with advanced compaction reduction technologies for off-road vehicles.

4.1 Tracks

Commercially successful track-type vehicles, which were recognized under the trademark name Caterpillar, began production in the early 1900s [27]. These early agricultural tractors, similar to the one shown in **Figure 8**, paved the way for future tracked vehicles and the continued use of the more complex metal grouser style tracks on construction equipment. Tracks did not remain popular in the agricultural sector, once pneumatic tires became available. They faded from use for many decades due to some specific issues that later rubber-belted machines were finally able to address.

While construction equipment is traditionally shipped to a worksite on a large trailer, tractors are generally driven from field to field on the road. Track-type machines with metal grousers are slower than pneumatic-tired machines during road transport under their own power. This slower transport speed, combined with a poorer ride for the operator and higher costs, eliminated most traditional track-type tractors from the agricultural market during the 1920s and 1930s. Two revolutionary designs, which are still produced by major manufacturers today, reintroduced the use of tracks on tractors. In 1986, Caterpillar launched the revolutionary rubber track Challenger 65® tractor, shown in **Figure 9** [3]. The Challenger used a two-track running gear system, similar to most construction equipment designs [28]. Shortly thereafter, Case IH introduced an articulated tractor with tracks at each corner of the machine. The Quadtrac®, shown in **Figure 10**, was configured like a traditional four-wheel drive tractor, with each of the contact points supported using a triangular track drive and bogie mechanism [29].

Figure 8.
Benjamin Holt testing the first prototype gasoline-powered track-type tractor in 1908 (Caterpillar, Inc., 2021) – [27].

DOI: http://dx.doi.org/10.5772/intechopen.104489

Figure 9.
A 1986 Caterpillar challenger 65® rubber-tracked tractor (TractorData.com, 2016) – [3].

Figure 10.
A 1997 Case IH Steiger Quadtrac® tractor (Case IH, 2022) – [28].

Today, rubber-belted tracks have become so successful that a common argument in the agricultural world is the debate of tracks vs. tires. However, opting for a tracked configuration creates a sizable increase in both purchasing and operating costs for the tractor. The price jump from tires to tracks can often be in the neighborhood of 10–25% of the cost of the machine. The operating costs jump too. As can be seen in **Figure 11**, the specific operational cost difference between tracks and tires for a 358 kW tractor is approximately \$0.085/kWh [29]. Currently, available data and existing published studies seem to support both sides of the tracks versus tires debate.

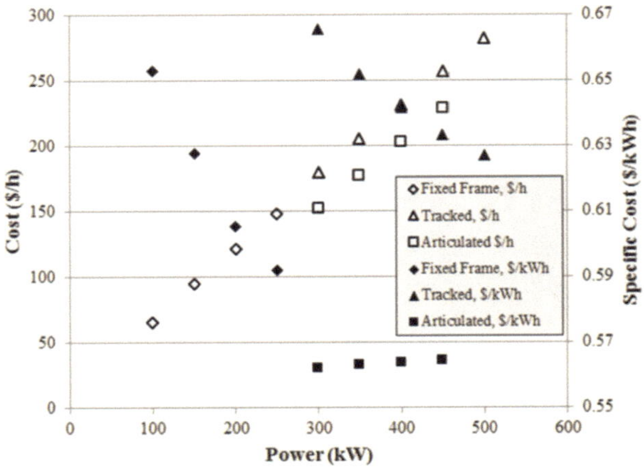

Figure 11.
Costs of operating tractors on tires and tracks (Case IH, 2022) – [28].

A 2018 European study involving a comparison of tracks and tires on two identical sugar beet harvesters revealed that the use of tracks does have a positive impact on reducing soil compaction [31]. Stress transducers were placed under the soil to analyze the compaction effects of the tractive devices. The mean ground pressure for the tire undercarriage system was measured to be 107 kPa, while the rubber tracks had a mean ground pressure of 84 kPa. As shown in **Figure 12**, ground pressure for the tires was also more concentrated, and it was more distributed under the tracks [31].

Because of their larger footprint, rubber tracks are often assumed to have a uniform weight distribution, but this is not true. Multiple design elements in a rubber track system, along with the integration of the track system onto the vehicle's frame, are critical to its effectiveness at reducing soil compaction. Common track systems, as shown in **Figure 13**, are traditionally composed of large driver and end wheels and smaller bogie wheels [29]. Bogie wheels, in theory, help to distribute the half the axle weight across the track's contact surface with the ground. However, in reality, the bogie wheels create ground pressure spikes beneath their relative positions. As can be seen in the right graph of **Figure 12**, the individual soil pressure peaks can be attributed to the bogies and wheels of the tracks evaluated in the study. The ideal

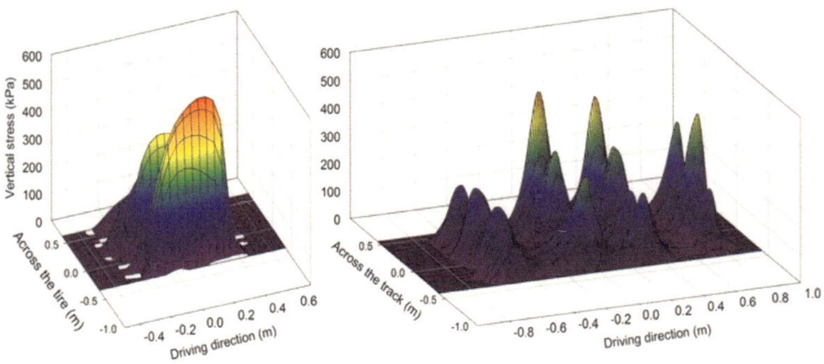

Figure 12.
Soil compaction study findings for a beat harvesting machine on tires (left) and the same machine on tracks (right) (Lamandé et al., 2018) – [29].

DOI: http://dx.doi.org/10.5772/intechopen.104489

Figure 13.
Bogie wheel track design (Case IH, 2022) – [28].

performance of a track can be identified by finding the theoretical applied ground pressure stress. This calculation is performed by dividing the load on the track by its contact area. In the sugar beet harvester study comparing tracks versus tires, the researchers discovered that the peak stress applied to the ground by the tracks was 5.7 times greater than the ideal ground pressure calculated value [31].

An analysis of the soil types across Europe was conducted to evaluate the maximum load capacity of different tractive devices, without causing permanent soil deformation [31]. **Figures 14–16** convey this analysis, showing soil types and the respective loads that can be handled by tires, tracks, and ideal tracks having a uniform pressure distribution. It is worth noting that a substantial load bearing increase could be achieved through improved track design.

Regardless of which side of the track versus tires argument is seen as the correct economic option, tracks do serve utility for farmers beyond that of tires. Although farmers prefer to be in the field when conditions are good, the weather does pose challenges. Depending on the geographic location of a farm and its soil type, it is common to deal with wet field conditions. As a result of the need to beat seasonal weather patterns, farmers often push acceptable limits to finish critical tasks in the field. Saturated soils are easier to tackle with tracks, because of their improved tractive performance over tires. Tracks are also less prone to rutting the soil in wet conditions. As seen in **Figure 10**, a side-by-side comparison of a tracked and tired machine shows the improved performance of tracks at staying on top of the soil. As can be seen, the track's footprint barely marks the ground, where the trailing tire cuts a deep rut. Severe soil compaction, like that caused by the tires in **Figure 17**, negatively impacts the health and performance of a farm's soil for the long-term [32]. However, the

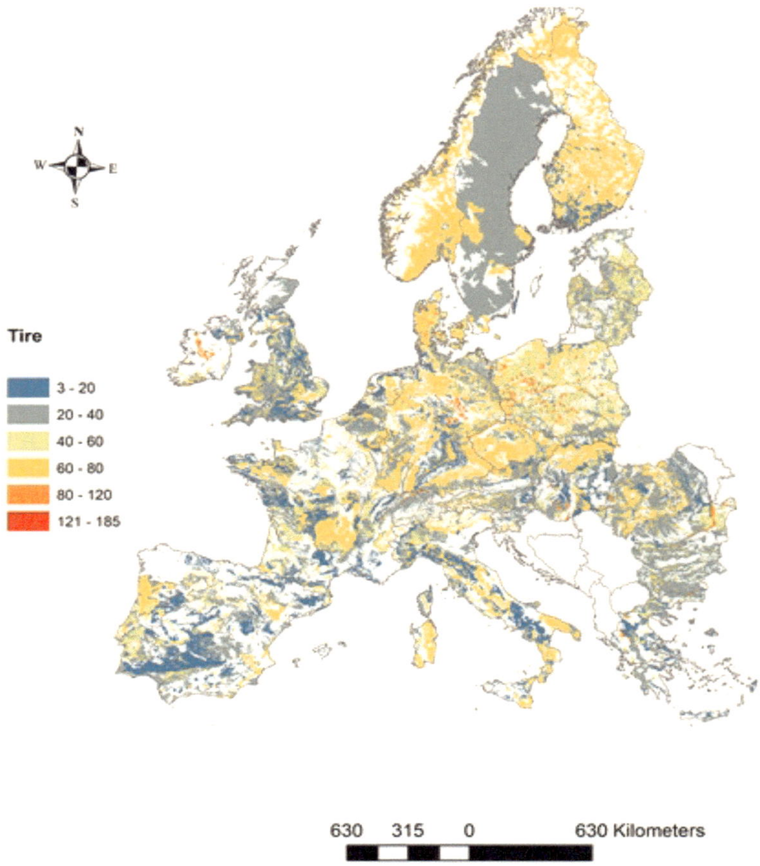

Figure 14.
European soil load carrying capacity map, showing the maximum load (kN) that can be carried by a 1050/50R32 tire without inducing permanent soil deformation at 0.35 m depth (Lamandé et al., 2018) – [29].

inability to harvest crops negatively impacts the economics and viability of a farm's business today. Although tracks are pricier than tires and may only provide limited benefits toward economically reducing soil compaction, tracks clearly outperform tires in adverse conditions.

4.2 Low inflation tires

Although the European study concluded that the use of tracks had a positive impact on soil compaction, differing studies have led to opposite conclusions [31]. The argument for tires is that correct maintenance needs to be performed to ensure that the tires are inflated correctly. A common issue is that farmers will over-inflate tires. Studies, like the one highlighted in **Figure 18**, reveal that incorrectly inflated tires create the most compaction [33]. In this specific experiment, correctly inflated dual tires were found to be impressively less compacting to the soil than tracks, while demonstrating that tracks could be superior to poorly maintained dual tires.

Tracks are undeniably an expensive, but great option for reducing soil compaction. However, low-pressure, properly-inflated tires are a potential option to match the benefits of tracks, at a fraction of the price. It is well documented that the depth of soil compaction is strongly correlated with tire pressure. Lower pressures cause less

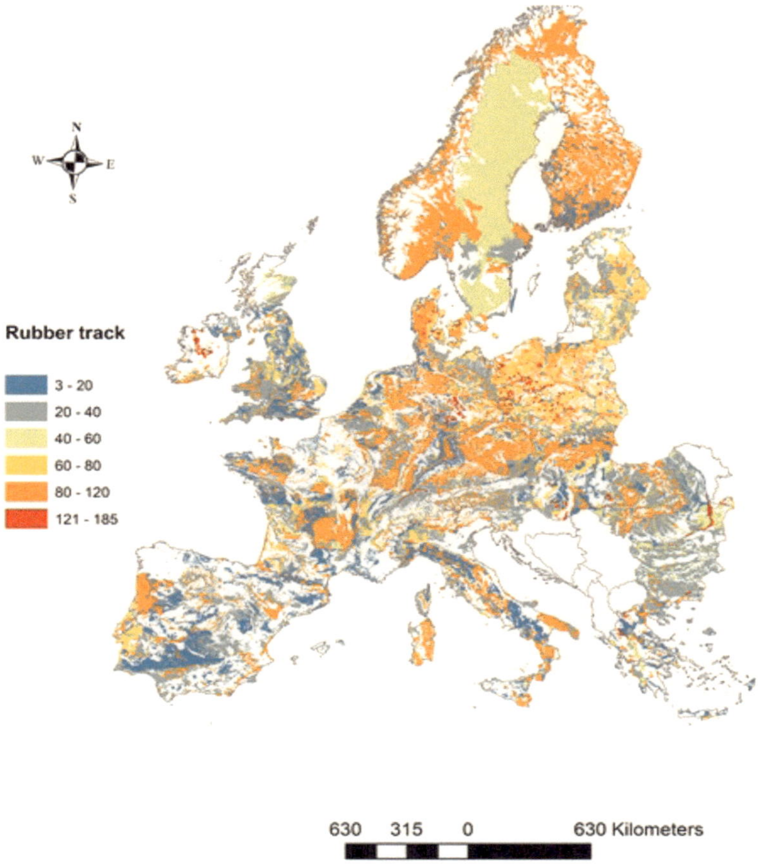

Figure 15.
European soil load carrying capacity map, showing the maximum load (kN) that can be carried by a rubber track without inducing permanent soil deformation at 0.35 m depth (Lamandé et al., 2018) – [29].

compaction. The limit to this practical method of reducing compaction is that the bead of the driving tires must remain on the rim. This low-pressure tire strategy helps reduce soil compaction by increasing the tire surface contact area with the ground. The increased contact area reduces the pressure exerted on the ground. Due to the limitations of decreasing the air pressure in traditional radial tires, tire companies have developed new flexion technology to allow even lower tire pressures. Increased Flexion (IF) and Very Increased Flexion (VF) tires, first introduced by Michelin during the 2000s, use a mature technology that greatly decreases the soil compaction from today's heavy machinery. The VF and IF tires can support the same loads with 40% and 20% less air pressure than radial tires, respectively by using increased tire sidewall strength [34].

While tires do not have as extensive of a surface area as most track designs, low pressure and Flexion-style tires make-up for some of the ground pressure shortcomings on tired vehicles, and in some applications, they can be a more viable option. Tracked vehicles experience pressure spikes at each bogie, whereas tires can be more consistent in the application of load to the soil [35]. Modern row crop tractors are commonly seen with dual rear tires and even dual front tires. As new equipment becomes larger and larger, single tires are no longer viable. As the demand for Modified Front Wheel Drive (MFWD) tractors has increased, additional weight has been

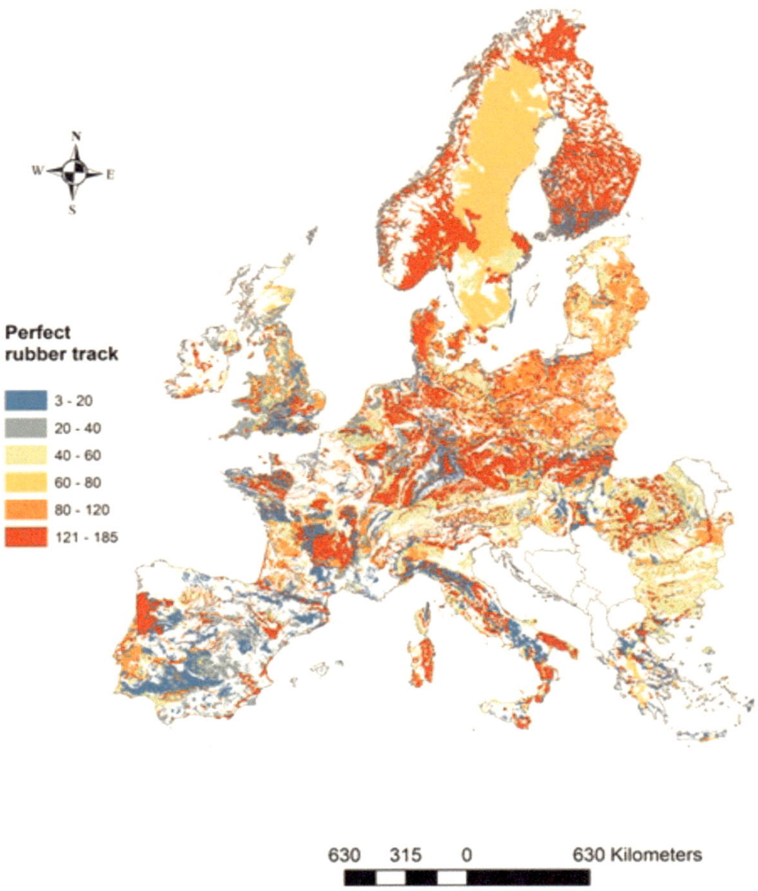

Figure 16.
European soil load carrying capacity map, showing the maximum load (kN) that can be carried by a rubber track with perfectly even stress distribution without inducing permanent soil deformation at 0.35 m depth (Lamandé et al., 2018) – [29].

added to the machines, requiring further soil compaction reduction methodologies to be undertaken. The most common MFWD tractors variants today have both dual front and rear tires. As would be expected with the addition of a second set of tires, soil compaction is reduced. This is achieved by essentially doubling the contact surface area [36]. The addition of a second set of tires allows for tire pressure to be reduced even further, also decreasing the potential for soil compaction [36]. These strategies can be combined for reasonably additive results. Using Flexion-type dual tires at low tire pressures can achieve even lower soil compaction. Under certain circumstances, properly inflated duals have been shown to be more effective at reducing soil compaction than tracks. Triple tires can be seen in certain high-power applications. However, they are not commonly seen in modern agriculture. The increased width of the tractor would be a benefit in the field, but transport on the road becomes infinitely more challenging. Axle stresses multiply significantly as well.

A recent innovation in agricultural tractor tires involves changing the overall design of the tires and the rim. New Low Side Wall (LSW) tires feature a significantly reduced tire aspect ratio, which results in a wider tire with reduced side walls. These LSW tires are intended to completely replace duals on modern farm equipment. While

Figure 17.
Rut comparison of tracks vs. tires in muddy conditions (Elmers manufacturing Inc., 2019) – [30].

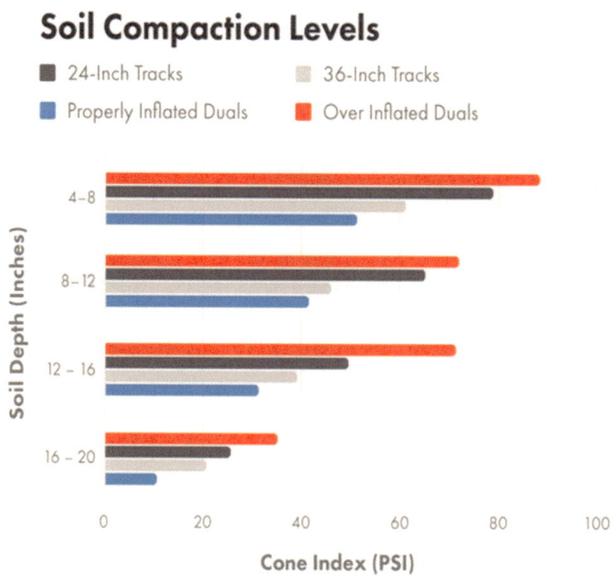

Figure 18.
Soil compaction comparison study findings (NTS Tire supply team, 2019) – [31].

these new tires are a more expensive initial investment, including a completely different rim and new tires, they are still cheaper than modern tracked systems. LSW tires could be a viable option for reducing soil compaction, as well as providing other benefits to the operator [37]. Tractors with a high center of gravity and LSW tires can experience reduced sway in motion, as well as better resistance to power hop [37]. LSW tires have a larger width allowing for more surface contact with the soil, as well as retaining the reduced inflation pressures similar to the Flexion-style tires [37].

Tires are typically the more attractive option for most farms, due to lower purchase and operational costs. Since the potential benefit of tracks is only for the reduction of soil compaction, tire systems, which can effectively compete with tracks in this metric, have a competitive advantage. With LSW tires closing the marginal difference in performance between tracks and tires, tires in many standard applications may be the smarter option. However, any option to reduce soil compaction will pay-out in the long run for farmers and growers. Conservation of the world's natural resources is imperative for the continued survival of humanity, especially with the extreme population growth projected for the next fifty years. Producing more food with less resource inputs is the goal of all of agriculture. Conserving the land is the first step toward a better tomorrow that will continue to be able to feed its people from the soil.

4.3 Two versus four wheel drives

Four-wheel drive vehicles can produce less soil compaction than their two-wheel drive counterparts, assuming all other factors are equal. Four-wheel drive systems also encounter less slip in motion and have a more optimal weight distribution, which likewise helps reduce the soil compaction. Slip can be thought of as a horizontal component of soil compaction. Slip occurs as the soil behind the tire compacts to support the drawbar load. There is a shrinkage in the matrix of the soil [38]. When traveling off-road, all vehicles have some amount of slip. This slip is determined by the interface between the wheels and the ground. The larger the ground contact area, the less slip occurs. Four-wheel drive vehicles have less slip than two-wheel drive vehicles. While two-wheel drive vehicles may have the same number of wheels on the ground, the non-driving wheels do not provide any traction. Tracked vehicles have an advantage as the entire length of the tracks are driven, and therefore, they have reduced slip when compared to tires. Nonetheless, slip sufficient to support the forward travel and drawbar loads on the machine still occurs. The reduction in soil compaction behind four-wheel drive vehicles has been demonstrated experimentally. **Figure 19** shows that the bulk density of soil was found to be 5.6% less than in rear wheel drive and 7.3% less than in front wheel drive vehicles [39].

From a practical perspective, four-wheel drive and two-wheel drive vehicles are built differently. Four-wheel drive vehicles are designed to have a different weight distribution. A rear-wheel drive vehicle has the center of mass at roughly one-third of the wheel base forward of the rear axle. A four-wheel drive vehicle has the center of mass located slightly more forward. This is advantageous, because the tractive force from the wheel depends on the normal force with the ground. Under drawbar load, the front and rear ends of the tractor are supported more equally, and the peak pressure on the ground is lower. Larger wheels have a higher area of contact with the ground, which results in lowered soil compaction. Many four-wheel drive vehicles have an articulated chassis used for steering. An articulated vehicle's axles follow only a single pathway when turning, which also reduces the area of compaction.

Just as certain soils are more prone to soil compaction, some soil types benefit more from four-wheel drive tractors. It is more difficult to gain traction in loose soil. As **Figure 20** shows, the moisture content in the soil plays a significant role in the compaction tendency of the soil [39]. Soils with a greater moisture content typically generate more slip [41]. As discussed earlier, slip is correlated with soil compaction. "Tire travel" will be significantly more in wet soil to cover the same distance.

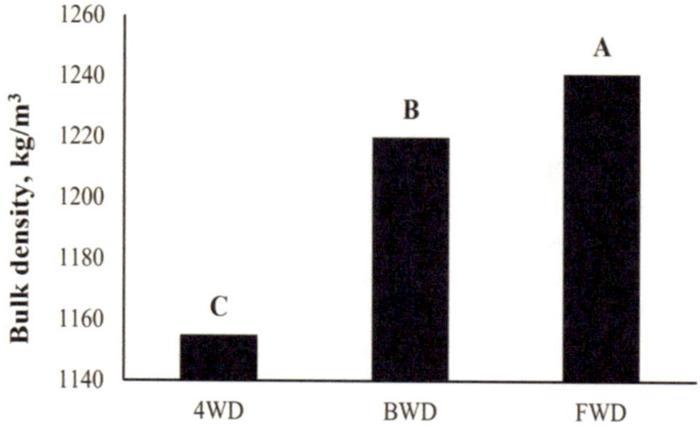

Figure 19.
Bulk density of soil following a tractor pass with different drive systems (Abu-Hamdeh et al., 1995) – [37].

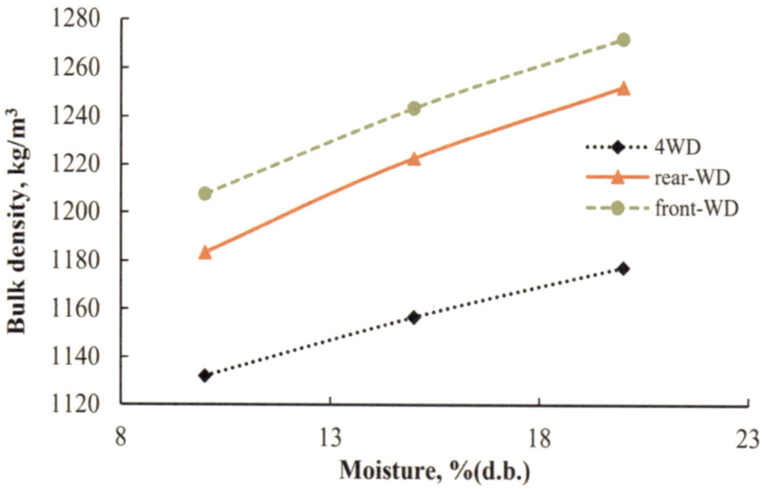

Figure 20.
Effects of moisture on soil compaction between multiple vehicles (Abu-Hamdeh et al., 1995) – [37].

4.4 Sensors, actuators, and special applications

Mechatronic agricultural systems are the future of agricultural machinery. One proposed means to reduce soil compaction is to utilize numerous smaller robotic machines, instead of progressively larger machines, to tend the fields. One limitation to further development along these lines is the price of a fleet of machines, while another is the human management factor. The price will likely come down as the technologies develop, but the human factor will remain stagnant until a "critical mass" of the new equipment enters the agricultural equipment market and demonstrates viability. These modern agricultural mechatronic systems will contain numerous sensors and actuators as their essential elements. Actuators perform the specific tasks directed by the vehicle's controller. Sensors facilitate the feedback from the actuators to the tractor's control system. The feedback data works as a performance measure for the actuators and the control system as whole. Specifically, the control system receives data on the success of the actuators' actions, the vehicle' position and motion, and the vehicle's immediate environment. Driveline control systems with the feedback mechanisms can successfully address soil compaction problems in many special applications. These automatic control systems have potential for use in the envisioned swarm systems for everyday agricultural operations. A swarm of smaller automated machines could become a disruptive technology, which would shift the paradigm in the current soil compaction reduction practices for crop production systems. The utilization of real-time feedback from soil conditions has multiple previous implementations for experience to be drawn from.

One example is a unique tractor for special climates. The Gidrokhod 49,061 ("Gidro" – hydro "khod" – traveler), shown in **Figure 21**, is "a three-axle all-wheel drive machine with a hydrostatic driveline with an automatic control system" [42], p. 147], which combines an individually driven axle design with a feedback-based approach to vehicle control. The Gidrokhod's driveline operates as follows: "[The] driveline is a full-flow mechanism that includes three axial-plunger controllable reversible and invertible hydraulic pumps and six axial-piston controllable and invertible hydraulic motors. Each pump is associated with two tandem hydraulic motors that set into motion the wheels of one hypothetical axle. The torques and rotational speed of the hydraulic motors are controlled individually by varying the displacements of the pumps and motors by means of an automatic control system" [42], p. 147]. The Gidrokhod's automatic control system supplies the required power to each wheel "as a function of the current conditions of interaction with the soil" [42], p. 147]. Gidrokhod was originally designed to reduce soil compaction from human activity in tundra, where the plants and soil are particularly vulnerable to any soil loading, such as the pressure from tracks or tires. The Gidrokhod's hydrostatic driveline also improves the vehicle's off-road drivability by dampening returned ground shocks into the driveline. The Gidrokhod's hydrostatic driveline is a computer-controllable, tested technology that could be transferred to off-road vehicle applications in agriculture to address the soil compaction problem.

Another example of a special off-road vehicle is a small off-world exploratory rover. These machines closely resemble hypothetical swarm agricultural vehicles and are essentially miniature space tractors. The pace of modern technology suggests that humanity will start colonizing the Moon and Mars by the mid-21st century. The

Figure 21.
Gidrokhod 49,061 (Vantsevich & Blundell, 2015) – [39].

comparison between an off-world exploratory robot and a small, swarm agricultural robot tractor is not as outlandish as it might first appear. The sensory apparatus necessary for independent wheel suspension control works as well minimizing soil compaction as navigating unknown terrain. The external manipulators resemble plant tending tools, and the ability to remain on-station and function unmanned is similar. It is conceivable that low-compaction inducing swarm agricultural tractors may look a great deal like our exploratory robots.

Russia was the first to send a rover to the Moon and Mars. While the Mars mission was a failure, the Soviet Moon exploration program laid the foundation for the future robotized space exploration missions. Over 50 years ago, in 1970, the Soviet Lunokhod 1 ("Luna" – Moon), shown in **Figure 22**, had a number of soil compaction sensors, a special wheel to measure traction, and single independent drives on each of its eight wheels for improved mobility [43]. As of now, multiple Mars rovers from the United States and one from China are traveling on the surface of Mars. Modern US rovers, like Perseverance, shown in **Figure 23** [44], combine the essential soil compaction sensors with the sophisticated modern drivetrain solutions, such as an advanced feedback loop from a complex sensor network, photo and video surveillance systems, as well as the use of Big Data concepts to better predict the ambient soil conditions and any possible action protocols during deployment and moving between operation areas.

The off-world researchers operating these rovers build terrain models on Earth, using transmitted data from the operating rovers, which has a 5 to 20 *min* signal delay. They use location information from satellites circling Mars, just like farmers on Earth do for agricultural production. Similar to how military location technologies came into the consumer world, space-based technologies will eventually find their best-use applications on Earth. One particular technology transfer path will be for the highly-accurate location technologies needed to control small robotic, low soil compaction-inducing vehicles for agricultural production. The first half of the 21st century will

Figure 22.
Lunokhod 1 moon rover (Kassel, 1971) – [40].

Figure 23.
Perseverance Mars rover (Wikipedia, 2021) – [41].

continue to see increasing technology transfer from electronics and space industry into everyday agricultural operations.

5. Soil compaction management in agriculture

Although the chassis and undercarriage design of tractors, combines, and harvesters is of obvious concern to engineers when trying to decrease soil compaction, agriculturalists have developed a variety of other methods and practices that contribute to the alleviation of soil compaction impact. These conservation tillage practices and alternative process design considerations are an important element in overall soil compaction reduction, as they can be applied to any and all farming operations, even those that do not have the most up-to-date equipment. Farm management practices can have a profound impact on reducing soil compaction, as well as maintaining soil organic content, reducing nutrient suppression, and decreasing time and energy spent in the field. From a sustainability standpoint, these conservation practices may even be more impactful for farm and field management at reducing soil compaction than any specific tractor or undercarriage design.

5.1 Tillage equipment and practices

The design of tillage equipment is an important and fruitful area of research for reducing soil compaction during the necessary ground-working operations. Tillage equipment design affects the ways in which tillage equipment interacts with the soil to help to alleviate long-term effects of disturbance in heavily-worked ground. Some of the core ideas within tillage implement design are load distribution, working point and shank design, working depth, the different types of soil disturbance, and the soil pulverization level. Most modern research is targeted at collecting specific information about the impacts of these conditions on soil health, compaction levels, and seed bed preparation, as well as energy use and the time spent in the field. The implications of tool design, structural loading, the types of conservation and reclamation equipment, and the impacts of soil compaction management on the energy consumption and overall performance of an agricultural venture will be examined in this section.

5.1.1 Tool design

In early agricultural practices, the moldboard plow dominated tillage as the most effective tool for turning the soil to create a seedbed. Its design was maintained in many forms of tillage equipment for years before its harmful impact on soil health, organic material, and erosion was realized. In contrast to the simplistic design of the moldboard plow, modern tillage equipment tool designs come in many shapes and sizes. The effects of tool geometry, orientation, depth, field speed, and other factors impact the level of soil disturbance and compaction. Various tool types can create a multitude of different outcomes in the upper soil layers in terms of soil aggregate size, topsoil density, porosity, and organic matter distribution. Other tools act predominately at the sub-surface level. In particular, deep cutting tines have the greatest impact on sub-soil compaction. Their shape, working depth, and spacing all affect the resulting soil compaction differently.

The effects of specific tine geometry and individual tine orientation were explored by researchers using finite element analysis (FEA) modeling [45]. **Figure 24** depicts the range of geometric variation explored, including the alteration of tine width, rake angle, and tilt angle. The primary results from this study concluded that at comparable field speeds, the increase in tine width linearly increased the resulting downward vertical force, while increasing rake or tilt angle linearly decreased the downward vertical force [45]. The implications of this study affect tractor power sizing, the uniformity of transmitted force along a vertical soil profile, the soil pulverization level, and the subsoil compaction. Most certainly, the results also present a variety of design trade-offs, depending on the immediate and long-term priorities of the specific farm manager. However, from the standpoint of reducing compaction whilst maximizing surface soil pulverization, minimizing the tine width and maximizing the tilt and rake angles create the least amount of sub-surface compaction.

It is important to note that the tilt angles can be non-uniform both on individual tines and on the overall tine set-up for an entire tillage unit. Many times, a compromise between minimizing draft forces, decreasing compaction, and managing soil upheaval can be achieved by applying a diverse range of different geometric and orientation values throughout a single tillage implement [46]. This becomes even more applicable the larger the implement is, due to the increased number of rows and columns of working points. Besides the considerations outlined above, two other vital components of tillage implement design are tool spacing and working depth in relation to the "critical depth". Critical depth is generally considered to be the point below which soil disturbances are concentrated near the working point and not distributed throughout the soil. **Figure 25** shows both the effects from operating below a critical depth and the dramatic increase in soil compaction as a result of tillage below this level [46].

Unfortunately, critical depth is not uniform by any means. It varies significantly with multiple variables, and it can be heavily impacted by moisture level, soil type, and the presence of a cover crop. This makes determining an operational depth a challenging task, particularly for inexperienced operators. Often initial passes are needed to estimate ideal working depths. There has been some research done regarding the use of strain gauges on subsoiler tines in conjunction with depth sensors,

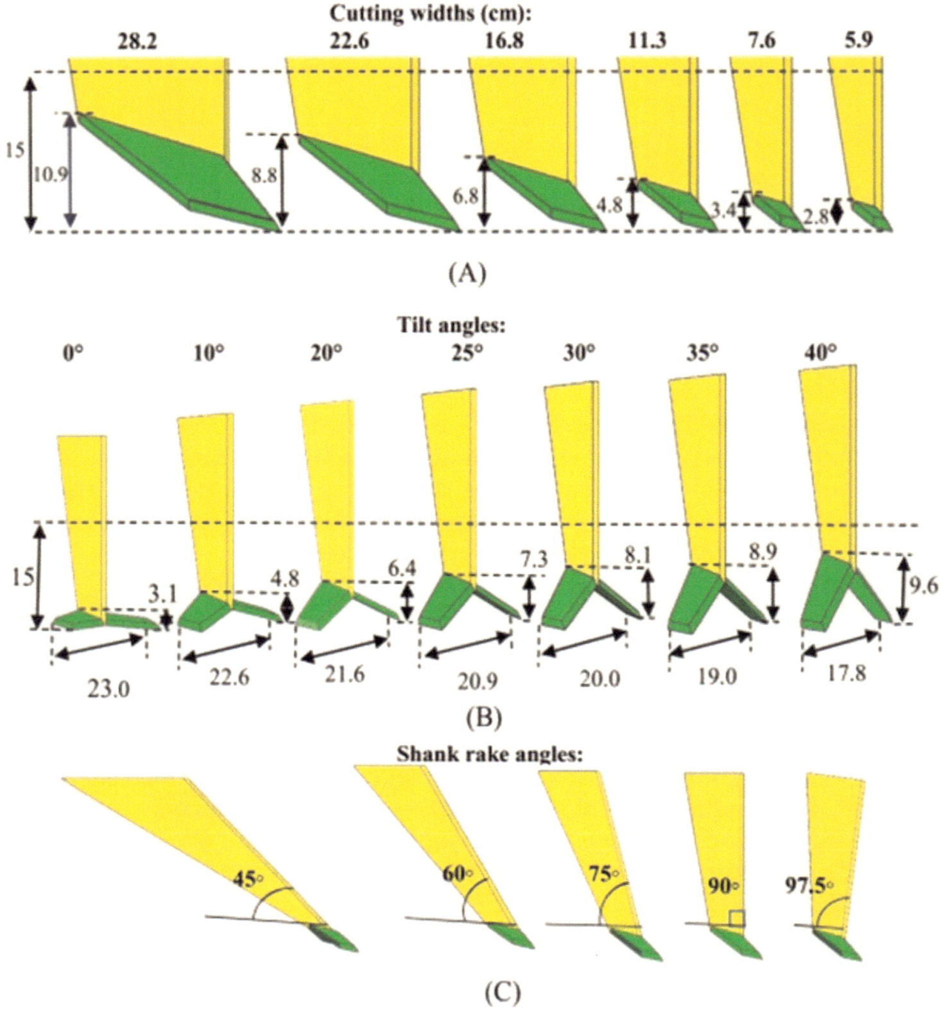

Figure 24.
(A) Six single sideway-share subsurface tillage implements with the same rake and tilt angles of 10° and 15° with different cutting widths; (B) dual sideways-share subsurface tillage implements with rake angle of 15° with different tilt angles; and (C) five dual sideways-share subsurface tillage implements with share tilt and rake angles of 10° and 15° with different shank rake angles (Hoseinian et al., 2022) – [42].

which can utilize a closed loop response system automatically adjusting height to maintain the desired draft and vertical forces [47]. These systems still require a degree of experience and skill to determine the expected shank loading at, above, and below the critical depth, in order to set the necessary system limits prior to operation. Although the practical difficulties with feedback-based systems are numerous, increased implementation of the above described depth adjustment mechanisms will provide a wealth of data regarding forces at and around critical depth. This information will only make these systems more effective in the future [48]. **Figure 26** illustrates the effects of tine spacing on overall soil disturbance. When tine spacing exceeds 1.5 to 2.0 times the working depth, an interesting phenomenon takes place, where the soil disturbance only occurs locally and results in a non-uniform subsoil profile and soil surface [46].

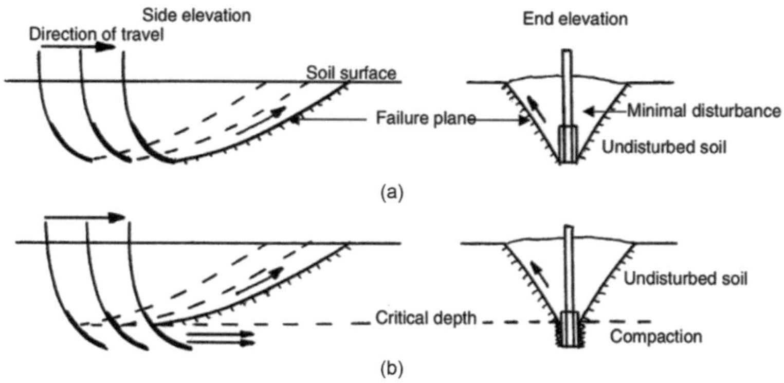

Figure 25.
Varying level of soil disturbance with narrow tine: (a) above critical depth; (b) below critical depth (Spoor, 2006) – [43].

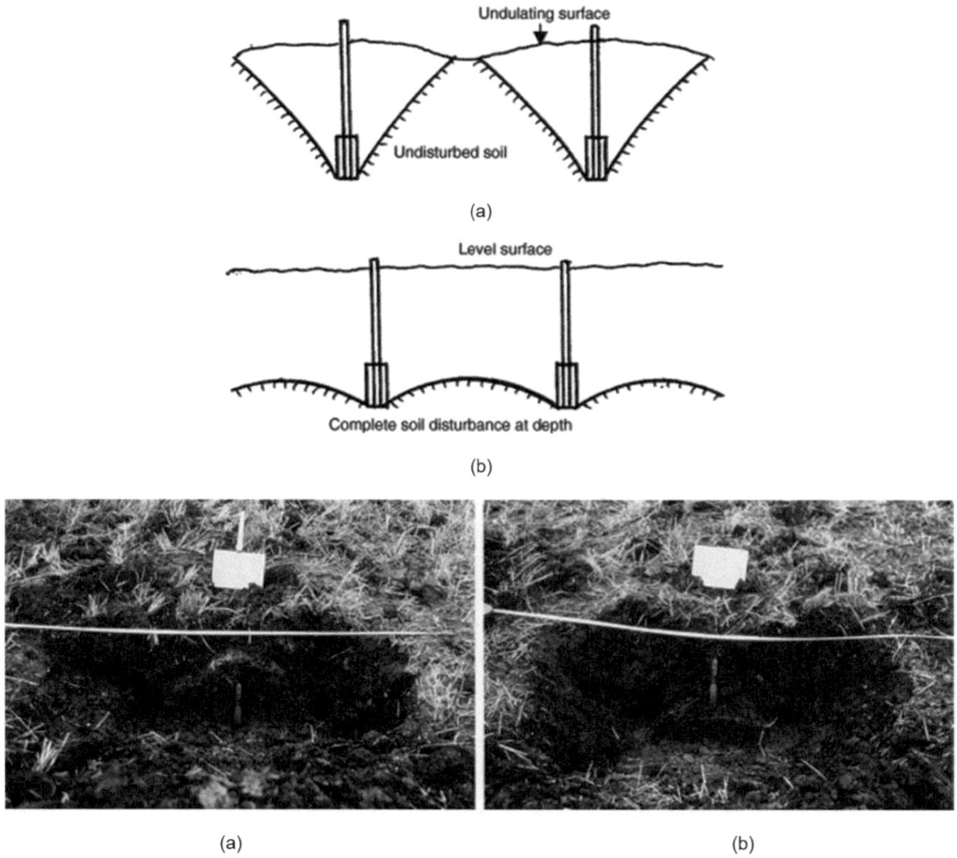

Figure 26.
Influence of tine spacing on the soil disturbance profile: (a) wide spacing; (b) narrow spacing (Spoor, 2006) – [43].

This outcome is likely to be troublesome for planting, as different row unit depth wheels will be penetrating the surface soil to different depths. The lack of consistency in seed depth, because of this poorly prepared seedbed, will result in emergence and germination issues. Although not initially obvious, the lack of uniform soil disturbance also affects compaction levels. Firstly, the lack of a uniform soil disturbance cross-section that occurs when using widely-spaced tines, illustrated in **Figure 19**-a, results in some subsoil being undisturbed. This soil remains compacted over time. When using a tillage implement with a wider tine set-up, it is easy for an operator to exceed the critical depth in order to achieve a cleaner surface profile, but in doing so, the subsoil compaction has been increased throughout the field. Using a narrow tine design dramatically decreases the chances that an operator will need to exceed critical depth in order to achieve the desired seed bed quality.

5.1.2 Structural loading

While magnitude of downward vertical force for tillage equipment simply does not compare to tractor units, it is still important to consider how the soil reacts with the implement loading as it moves through the field and what factors play into determining the optimal number of tines and the structure of tillage equipment. There are three primary ways in which soil reacts to the loads and forces placed on it by cultivation implement: brittle loosening disturbance, compressive disturbance, and tensile disturbance [43]. Brittle loosening occurs when the implement load compresses the soil and causes a sliding or slipping during the operation. The effects of the sliding and slipping are such that the soil aggregates, clumps, and masses move relative to one another. The overall volume of soil masses is increased, cracked, and spread-out. Contrary to compressive disturbances, a large quantity of the soil is actually decompressed or loosened as a result of brittle loosening. This is the kind of soil response that occurs primarily under ideal loading and working depth conditions.

Compressive disturbance also occurs under compressive loading, but without the exposure to masses sliding relative to one another. In this case, without sliding, the soil is more likely to experience high degrees of compression and increases in density. This process is more common using heavier implements, when there is a low draft force. Tensile disturbance is virtually the same as brittle loosening and has similar results, such as decreased density and alleviated compaction. The difference lies in the fact that tensile disturbance occurs when soil aggregates are pulled-away from one another and forced to spread-out. This kind of disturbance is more likely to occur under high moisture conditions, where the load is cushioned and absorbed to a greater extent, thus negating the compressive impact of the load.

Each of the three kinds of soil matrix disturbances can be modified and impacted by the working depth, operation speed, and the weight of the implement. **Table 1** provides the basic tendencies for determining the design of the implement, based on the power of the tractor unit, and for potentially determining necessary engine power or anticipated working depth, based on the tine and structural design of the tillage implement and its working points. **Table 1** can be used for reclamation projects in which the soil has experienced long-term compaction and where aggressive subsoiler action is needed to prepare the soil for further tillage and planting preparation [46].

5.1.3 Soil loosening equipment

There is a big difference between common tillage equipment used for routine crop cultivation, associated with planting and harvest, and machinery used to rejuvenate the soil from excess compaction. Robust subsoilers are utilized when efforts are made to restore long-term compacted soil. These subsoilers must be capable of decreasing soil

Tractor size		Capability	
Engine power (hp/kW)	**Ballasted weight (tonnes)**	**Working depth range (cm)**	**Number of tines**
30/23	1.5	20–30	1
60/45	3.0	30–40	1
75/56	3.75	35–45	1
		25–30	2
100/75	5.0	40–50	1
		30–35	2
		25–30	3
125/95	6.25	45–55	1
		35–40	2
		25–30	3
150/110	7.5	50–60	1
		35–45	2
		30–35	3
		25–30	4
200/150	10.0	40–50	2
		35–40	3
		30–35	4
		25–30	5
250/185	12.5	45–55	2
		40–45	3
		35–40	4
		30–35	5
		25–30	6
For crawler tractor in same horsepower range, increase number of tines by 50% or working depth by 20%.			

Table 1.
Wheeled tractor capability for operating loosening tines in compacted soil (Spoor, 2006) – [46].

density and effectively disturbing the mid-subsoil level to make the land workable under normal cultivation protocols. As seen in **Figure 27**, these reclamation subsoilers typically utilize a three-point hitch attachment for depth adjustment, rather than a drawbar attachment and trailing configuration [46]. One issue with these subsoilers is the need to operate below the critical depth to create an adequate soil disturbance to restore the soil profile. Unfortunately, this process can cause further, deeper subsurface soil compaction, despite alleviating the compaction in the upper subsurface soil levels.

5.2 Controlled traffic farming

Since compaction is inevitable in agricultural operations, its minimization through operational management is critical to long-term sustainability. The essential principles of compaction management are the reduction of both tillage and field traffic. Modern

Figure 27.
Reversible subsoiler and its impact (Spoor, 2006) – [43].

best practices decrease these elements in crop production processes to the smallest feasible levels. Compaction mitigation techniques are reviewed in this subsection.

5.2.1 Low-till

The first management practice that can be used to reduce soil compaction is the low-tilling method. There are several aspects to low-till that help reduce erosion and soil compaction collectively. Low-till involves planting with a seed drill after a minimally-disturbing tillage operation. The soil is not as exposed to and penetrated by wind and water under this protocol. Low-till keeps an estimated 30% minimum of crop residue on the soil surface. This allows for more organic material to remain present in the topsoil, increasing the soil stability [45]. With low-till, water erosion is inhibited, due to the higher surface trash coverage and lower general depth of water penetration. Low-till farming protocols are an extremely popular choice currently, creating a nice compromise between conventional agricultural practices and more extreme conservation processes.

5.2.2 No-till

No-till farming is extremely effective at helping soil health in multiple different ways. With this method, only the soil surrounding the seed trench is tilled by the row crop planter. No additional tillage operations are performed. Besides being extremely cost effective in fuel consumption, no-till operations have very quick positive results, when compared to other methods. In as short as 2–5 years, soil compaction will naturally be reduced in the topsoil, as microorganisms and organic material increase and expand in the soil. The increased biomass will have a longer lasting effect, as the crushing strength of the soil will be dramatically increased. In clay soils, these results may be more pronounced. Compacted clay soils create the tightest restriction of all soil types. Allowing for root penetration and added biomass expands clay soil until it is much less compactable. With the no-till method, the higher vegetative density alone can help absorb impact from smaller implements. With the proper planting equipment, the no-till method is a very simple and effective method for reducing and reversing soil compaction [49].

6. The causes of soil compaction

Soil compaction is the phenomenon associated with the collapse of soil media to support the loads imposed upon it. All agricultural operations on the surface of the ground cause soil compaction. Heavy axle loads, wet soil operations, livestock grazing, and materials stored directly on the surface can all result in unwanted compaction. The details of these agricultural process root causes of soil compaction will be explored in this section.

6.1 Operation of equipment with heavy axle loads

An axle load is the total load supported by a single axle, usually across two points of contact on either side of the vehicle. Although most agricultural equipment uses two axles for load distribution, each point of contact carries harmful loads into the soil. A large agricultural vehicle weighing 20 *ton*, creates 10 *ton* of force on each axle and causes the soil beneath each point to compact, until it can support the imposed load. The biggest factor to consider in reducing soil compaction is large axle loads. For two vehicles with the same weight distribution, the bigger the vehicle's contact area with soil, the lesser the pressure is applied to the soil surface. **Figure 28** illustrates an advantage of tracks over tires by the contact area parameter [22]. Research has shown that having an axle load of 10 *ton* can cause deep (more than 45 *cm*) subsoil compaction under moist conditions [8]. Grain carts and other heavy trailing implements behind the power units add to the problem of soil compaction, since axle load is determined by the total weight of the vehicle divided by the number of axles. Reducing single axle loads below five *ton* or less will diminish subsoil compaction, and only cause topsoil compaction [8]. Using heavy machinery under wet or moist conditions always increases soil compaction dramatically over use under dry conditions for most

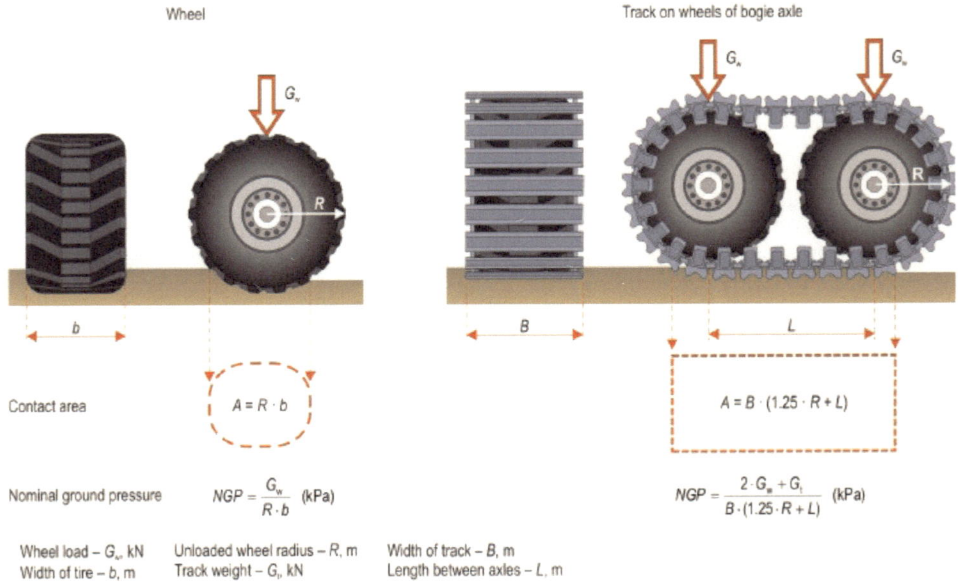

Figure 28.
Tracks versus tires load distribution areas (Mellgren, 1980) – [22].

soil types [23]. The relationship among pressure applied, water content and bulk density varies across different soil types as particles rearrange with changing water contents [24].

6.2 Operation during non-optimal soil conditions

Under non-optimal soil conditions, field farm operations should be considered with great reluctance, due to the potential for severe damage to the soil matrix. As farm equipment crosses through a wet field, ruts are formed from soil compaction around the tire path. Tillage is a common practice to relieve soil compaction due to poor soil management. However, tilling breaks-apart the soil structure and causes further traffic, in addition to deeper compaction in the field. A tilled soil is more easily compacted, since the subsoil beneath the tillage line is now in a more vulnerable state for soil compaction [25]. Under good soil conditions, the integrity of the soil is reasonably strong and minimizes the loss of pore space from heavy equipment travel. When soil conditions are non-optimal, the structural integrity of the soil is significantly reduced, and this results in the elimination of pore space with vehicle traffic. As shown in **Figure 29**, when the same pressure is applied in a loam soil, the bulk density significantly increases with increasing soil water content, thus, leaving the soil susceptible to compaction [24]. Additionally, water within the soil matrix reduces the coefficient of friction between neighboring soil particles and promotes the ease of displacement and flowability of the soil.

6.3 Livestock grazing

Livestock grazing can affect soil stability and functionality if not managed properly. The severity of soil damage due to livestock grazing is related to the soil type, texture, and moisture content. Pugging, the formation of soil around the hoof of the

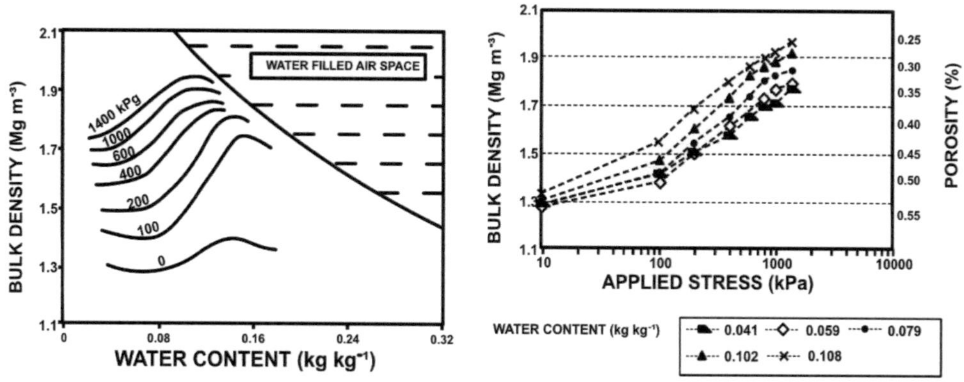

Figure 29.
Water content, pressure applied and bulk density diagram (left) and compression curve for a loam – Typic Haplaquept soil (right) (Smith, Johnston, & Lorentz, 1997) – [23].

livestock, can result in increased soil compaction and a reduction in soil surface water infiltration rates [26]. When water does not infiltrate through the soil surface during rainfall or irrigation, puddling occurs in fields. The trampling and pugging from livestock onto soil surfaces damages the subsurface soil integrity. The density of the livestock per unit of area in a pasture impacts the level of soil compaction due to pugging. This effect also negates the value of winter grazing on crop land to glean harvest losses. The long-term damage from soil compaction to the crop ground greatly outweighs the value of the "free" feed gained.

6.4 Other

Aside from intensive farming and grazing practices common in modern agriculture, there are other factors, some environmental and some man-made, that can have a noticeable effect on soil compaction. Depending on the region of agricultural production, the type of soils, as well as natural and artificial drainage, some fields can be subject to prolonged ponding of water in localized areas. Over time, the weight of the water ponded on the soil surface causes the soil pores to collapse further, slowing the movement of water through the soil and increasing the weight of water on top of the soil surface during future precipitation events. Water ponded on the soil surface adds 10 *kPa* of pressure per m of depth. Additionally, slowed water movement through the soil increases the risk of farming operations occurring during non-optimal soil conditions. Another non-conventional contribution to soil compaction is the relatively new practice of storing grain in large plastic bags that are laid-out on the soil surface. Producers using this method of temporary grain storage have noted significant soil compaction on the surface due to the weight of the grain.

6.4.1 Dedicated tramline equipment

The newest realm of controlled traffic farming incorporates unified implements that minimize in-field travel in a variety of ways. NEXAT GmbH is a leader in this field. The company has developed a single equipment carrier, known as a beam tractor, capable of planting, soil cultivation, crop treatment, and harvesting. They refer to this as the NEXAT System [50]. This fascinating piece of equipment, pictured in **Figure 30**, manages to minimize the required crop production machinery, is fully integrated, and does not require additional equipment or chassis components. It can keep-up with the advancing digital age of electronic controls and even has autonomous guidance. However, its most impressive feature is an ability to reduce the land driven-on from 60 to 80% to less than 5% by only traveling on dedicated drive lanes. NEXAT-like systems are crucial to the continuing effort of reducing soil compaction through the minimization of machinery footprint on arable land.

6.4.2 Tillage timing

Even the simple aspect of the timing of the tillage in a field can play a major factor in soil compaction. Early season tillage is often performed to reduce the weed density late in the season. However, early season tillage often is the wrong choice for both soil compaction and weed control during the growing season. Late-season tillage allows for more organic material to be added to the soil, while actively and drastically reducing the number of weeds present in the crop's growth cycle. Early tillage during the wet spring times increases the soil's tendency toward compaction. Heavy equipment and traffic through the fields amplify the destruction of the soil's internal structure. Decreased pore space and limited soil and water volume can result from wet soil tillage during the early parts of the crop production season [51].

The impact of tillage operations during non-optimal, wet conditions is a common concern for farm managers, and research into the actual implications of these kinds of operations is rather common. **Figures 31** and **32** below detail the results of a study looking into the change in resistance to soil penetration following tillage during wet conditions and the progression of the soil aggregate strength throughout the growing season for these soils [52]. **Figure 26** shows that after non-optimal cultivation, penetration resistance increased slightly compared to a reference soil that was not tilled but that this resistance was still significantly lower than heavily compacted soil. The true consequences of non-optimal tillage operations are exposed in **Figure 27**, in which it is demonstrated that the tilled soil is unable to recover during the following growing cycle. As a result, the tilled soil maintains a very high soil aggregate tensile strength

Figure 30.
NEXAT system for controlled traffic farming (Misser Uitgeverij B.V., 2021) – [47].

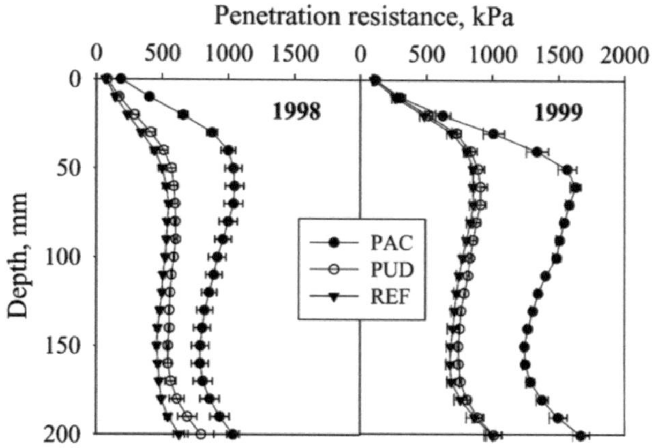

Figure 31.
Soil penetration resistance measured shortly after tillage operations in may 1998 and 1999. PAC: Compacted; PUD: Intensive rotary cultivation (Munkholm & Schjonning, 2004) – [49].

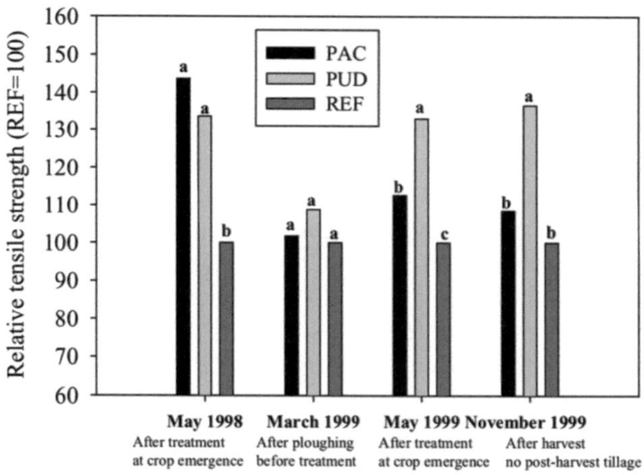

Figure 32.
Relative tensile strength (REF = 100) of air-dried soil aggregates (average of the four size fractions) at the different times of sampling. PAC: Compacted; PUD: Intensive rotary cultivation (Munkholm & Schjonning, 2004) – [49].

over time, further decreasing the soil's productivity and the long-term sustainability of agricultural operations in such soil.

6.5 Cover-cropping and crop rotation

The final aspects of farm management that impact soil compaction and soil health are the decisions that farm managers make regarding crop rotation and cover cropping. Both have specific impacts for nutrient availability and storage, organic material availability and control, weed control, and erosion prevention. However, both cover cropping and crop rotation can also impact the prevention of soil compaction. This section will review the impacts of cover cropping versus crop rotation, an outline cover crop selection to achieve maximum compaction prevention and maintain the necessary levels of erosion prevention, and the impact of pre-planting cultivation and its effects on seed bed, germination, and root development.

6.5.1 Cover-cropping vs. crop rotation

Cover-cropping is the practice of planting legume and grass varieties after the primary harvest, in the late fall, winter, or early spring before planting. Typically, these cover-crops are planted to instill nutrients into the soil, increase the organic material in the topsoil layer, and to better hold the soil together during tillage to prevent erosion issues. In addition to promoting yield advantages, cover cropping can also be used to improve the soil profile and decrease existing compaction through the creation of pores and reduction of soil bulk density.

Crop rotation aims more at cycling specific nutrients within the soil matrix to promote a greater yield for specific crop types during different cyclic years. A good example of this is the common corn and soybean rotation, in which soybeans are rotated-in, when soil nutrient sampling indicates low nitrogen levels. Soybeans are utilized in this way, due to their nitrogen fixing attributes. This locks excess atmospheric nitrogen beyond what is needed for the soybean crop into the soil, to be used by corn in the following years. Crop rotation can additionally impact topsoil and subsurface soil compaction, because of the differences in root penetration profiles. This can aid in moisture uptake and retention.

One study looked at the difference between cover-cropping and crop rotation and then compared the impact on yield results, as well as the resulting soil nutrients [53]. The findings were such that in the short term, there was little evidence to say that cover-cropping alone could result in an adequate yield improvement, but a combination of cover-cropping and crop rotation promoted increased crop yields and retained the benefits of using cover-crops. On the other hand, when examining the effects on soil compaction, the long-term consequences of cover-cropping helped to dramatically negate long-term compaction issues. Cover-cropping plays an essential role in decreasing soil compaction through the reduction of soil bulk density, the alteration of soil aggregate size, the creation of root channels, and improving the aeration and pore space within the soil. Specific cover-cropping can also help to combat long-term compaction by promoting subsoil disturbances via root channels.

6.5.2 Cover-crop selection

One of the primary ways in which cover crops can impact soil compaction is through the creation of pore space and root channels. These openings help to decrease the soil's bulk density, break-up previously compacted volumes, and promote water infiltration, all of which further aid in this endeavor. **Figures 33** and **34** depict the results of studies on the effects of root profiles and root penetration resistance, which indicate compaction relief from cover-cropping [54, 55]. In particular, the studies investigated the differences in channels created by soybean and canola plant roots, as well as the effects on soil nutrient and water content from a variety of other legume-type cover crops [54, 55]. Cover-cropping with radish and legume type crops aided in decreasing the soil penetration resistance during later planting, and it marginally disrupted soil compaction. In addition, cover-cropping had added benefits for nutrient content and water availability. The data from WREC in **Figure 34** showed how cover cropping impacted soil with historically high compaction [55]. Utilizing cover-crops with large root profiles was particularly effective at increasing the macroporosity and facilitating aggregate break-up in both topsoil and subsoil [54]. The latter is particularly important for soil types with increased risk of compaction, such as those with a high clay content. In addition to its other benefits, cover-cropping is a useful

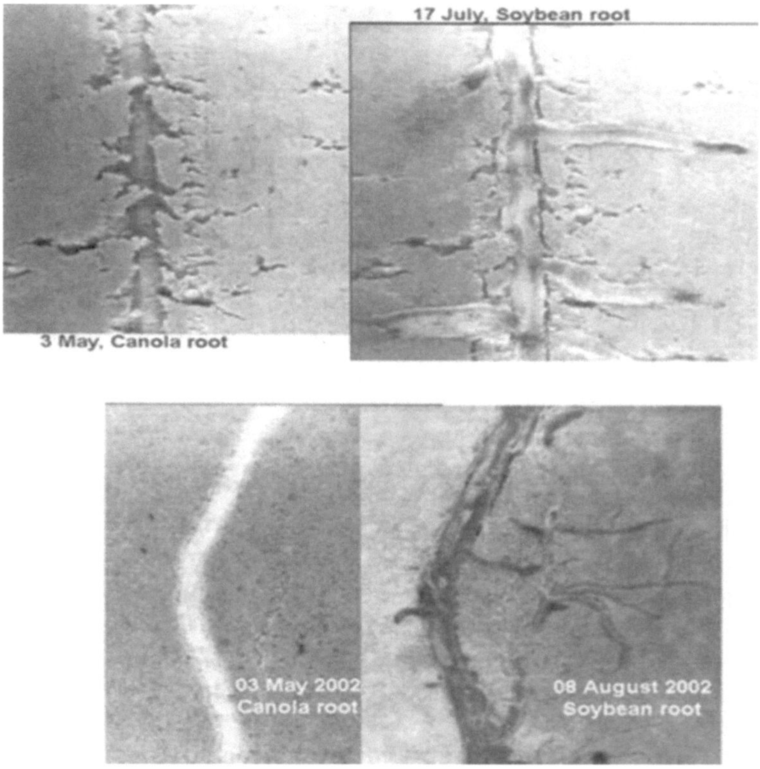

Figure 33.
Minirhizotron images showing canola roots growing in may (left) and soybean roots observed in July and august (right) following the channels made by the preceding canola cover crop at 38.2 cm (at WREC) (top) and 18 cm (at BARC) (bottom) depth. The bulk density was 1.55 and 1.61 g/cm^3 and penetration resistance was 2247 and 2176 kPa for the upper and lower soils, respectively (Calonego et al., 2017) – [51].

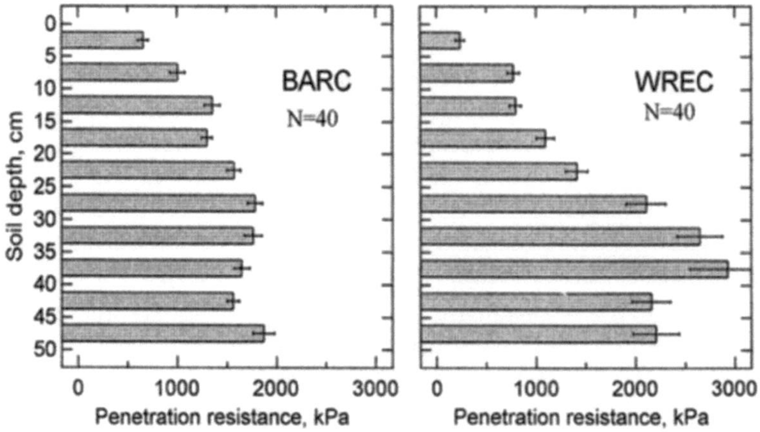

Figure 34.
Penetration resistance (kPa) with depth (cm) at Beltsville agricultural research Center (BARC) and wye research and education Center (WREC). The average volumetric water content at time of penetration resistance measurement was 0.22 cm^3/cm^3 (WREC) and 0.27 cm^3/cm^3 (BARC) in the surface soil (0–20 cm) and 0.29 cm^3/cm^3 (WREC) and 0.39 cm^3/cm^3 (BARC) in the subsoil (20–40 cm) (Williams & Weil [56]) – [52].

and inexpensive tool to aid in alleviating the effects of previous compaction, costing far less than mechanical relief applied through subsoiling operations.

7. Conclusion

This chapter provided a comprehensive review of the soil compaction problem. The historical aspects, the mechanism, and the environmental implications of soil compaction were discussed. A large off-road vehicle maneuvering with a draft load causes the soil beneath each point to compact, until it can support the imposed load in both the vertical and horizontal directions. Different vehicle designs have advantages and disadvantages in addressing the soil compaction problem. The track versus tires debate continues to this day, and the farmer's choice should depend on their specific situation. Sophisticated farm management practices can significantly reduce soil compaction in the mid-term. Farmers risk losing 50% of their expected profits, when the soil compaction is not addressed in a sustainable way. Farmers and policymakers are encouraged to work toward reducing and reversing soil compaction for sustainable management of agricultural lands.

The issue of soil compaction is not one that will ever cease to exist. It will continue to cause trouble for those in agriculture, construction, mining, and other industries that deal with soil and ground working. As such, it is important that an understanding of the impact of soil compaction is continually being disseminated into these industries, as well as the basic management practices that can help to prevent an extensive spread of the problem. For design engineers in these fields, soil compaction offers the potential for the continued improvement in equipment design. Looking specifically at agriculture, the on-going trend of increasing equipment size and capacity in order to improve fuel and energy sustainability indicates that there will be a continued demand to further reduce the equipment loading and improve soil interaction of crop machinery, in order to maintain an adequate level of soil compaction minimization. The design of tillage equipment has already come a long way from the moldboard plow, specifically in terms of minimizing soil disturbed unnecessarily, while maximizing the implement's capacity to pulverize the soil aggregates within the seed-bed. Moving forward, tillage equipment's most likely challenge will be ensuring adequate wheel support during operation, without causing additional soil loading and compaction forces.

Farm managers must recognize that the prevention of unnecessary soil compaction is of paramount interest in the long-term productivity of their resources. They need to adopt a continuous improvement attitude and do whatever is feasible to minimize compaction. The seemingly small benefits of tillage cycling, crop rotation, and cover cropping should not be overlooked, since these practices continue to prevent soil compaction and reduce equipment traffic in the fields. As with many other aspects of off-road vehicle and machine design, committing to improving the performance of all factors will increase the effectiveness of the soil compaction control and prevention areas. Because the ability to sustainably grow food is critical to humanity's future, agricultural engineers of the 21st century with a working knowledge of soil compaction phenomena will continue to be in high demand.

Acknowledgements

We would like to acknowledge our fellow classmates from the Fall 2021 Design of Off-Road Vehicles class at Purdue University's School of Agricultural and Biological

Engineering for their contributions to the structure and content of this technical chapter. Robert M. Stwalley IV is acknowledged for his work on the figures within the document.

Conflict of interest

The authors declare no conflict of interest.

Author details

Michael M. Boland[1], Young U. Choi[1], Daniel G. Foley[1], Matthew S. Gobel[1], Nathan C. Sprague[1], Santiago Guevara-Ocana[2], Yury A. Kuleshov[2] and Robert M. Stwalley III[1*]

1 Purdue University Agricultural and Biological Engineering, West Lafayette, Indiana, USA

2 Purdue University Polytechnic Institute, West Lafayette, Indiana, USA

*Address all correspondence to: rms3@purdue.edu

References

[1] Lane M. The Story of the Steam Plough Works : Fowlers of Leeds. Havertown, PA: Northgate Publishing Co.; 1980

[2] Raney WA, Edminster TW, Allaway WH. Current status of research in soil compaction. Soil Science Society of America Journal. 1955;**19**:423-428. DOI: 10.2136/sssaj1955.03615995001900040008x

[3] TractorData, "Challenger 65," 2016. [Online]. [Last Accessed 29 November 2021].

[4] Hadjilambrinos C. Reexamining the Automobile's past: What were the critical factors that determined the emergence of the internal combustion engine as the dominant automotive technology? Bulletin of Science, Technology, and Society. 2021;**41**(2–3):58-71. DOI: 10.1177/02704676211036334

[5] Soane BD, van Ouwerkerk C. Soil compaction problems in world agriculture. In: Developments in Agricultural Engineering. Vol. 11. Amsterdam: Elsevier B.V; 1994. pp. 1-21. DOI: 10.1016/B978-0-444-88286-8.50009-X

[6] Spence CC. God Speed the Plow: The Coming of Steam Cultivation to Great Britain, Urbana. IL: University of Illinois Press; 1960

[7] Horn R, Domzzał H, Słowińska-Jurkiewicz A, van Ouwerkerk C. Soil compaction processes and their effects on the structure of arable soils and the environment. Soil and Tillage Research. 1995;**35**:23-36. DOI: 10.1016/0167-1987(95)00479-C

[8] Duiker S. "Avoiding soil compaction," 2005. [Online]. Available: https://extension.psu.edu/avoiding-soil-compaction#:~:text=Compaction%20in%20the%20topsoil%20can,acre%20actually%20traveled%20are%20recommended. [Last Accessed 21 November 2021]

[9] Batey T. Soil compaction and soil management - a reveiw. Soil Use and Management. 2009;**25**(4):335-345. DOI: 10.1111/j.1475-2743.2009.00236.x

[10] Botta GF, Tolon-Becerra A, Bienvenido F, Rivero D, Laureda DA, Ezquerra-Canalejo A, et al. Sunflower harvest: Tractor and grain chaser traffic effects on soil compaction and crop yields. Land Degradation & Development. 2018;**29**:4252-4261. DOI: 10.1002/ldr.3181

[11] Soane BD, van Ouwerkerk C. Implications of soil compaction in crop production for the quality of the environment. Soil and Tillage Research. 1995;**35**:5-22. DOI: 10.1016/0167-1987(95)00475-8

[12] Torbert HA, Wood CW. Effects of soil compaction and water-filled pore space on soil microbial activity and N losses. Communications in Soil Science and Plant Analysis. 1992;**23**:1321-1331. DOI: 10.1080/00103629209368668

[13] Botta GF, Tolon-Becerra A, Tourn M, Lastra-Bravo X, Rivero D. Agricultural traffic: Motion resistance and soil compaction in relation to tractor design and different soil conditions. Soil and Tillage Research. 2012;**120**:92-98. DOI: 10.1016/j.still.2011.11.008

[14] Shaheb MR, Venkatesh R, Shearer SA. A review of the effect of soil compaction and its management for sustainable crop production. Journal of Biosystems Engineering. 2021;**46**:

417-439. DOI: 10.1007/s42853-021-00117-7

[15] Voorhees WB. The effect of soil compaction on crop yield. SAE Transactions. 1986;**95**(3):1078-1084. Available from: https://www.jstor.org/stable/44725467

[16] Jamali H, Nachimuthy G, Palmer B, Hodgson D, Hundt A, Nunn C, et al. Soil compaction in a new light: Knowing the cost of doing nothing - a cotton case study. Soil & Tillage Research. 2021;**213**: 105158. DOI: 10.1016/j.still.2021.105158

[17] Chamen T. Controlled traffic farming - from worldwide research to adoption in Europe and its future prospects. Acta Technologica Agrriculturae. 2015;**18**(3):64-73. DOI: 10.1515/ata-2015-0014

[18] Zabrodskyi A, Sarauski E, Kukharets S, Juostas A, Vasiliauskas G, Andriusis A. Analysis of the impact of soil compaction on the environment and agricultural economic losses in Lithuania and Ukraine. Sustainability. 2021;**13**: 7762. DOI: 10.3390/su1347762

[19] Gunjal K, Lavoie G, Raghavan GSV. Economics of soil compaction due to machinery traffic and implications for machinery selection. Canadian Journal of Agricultural Economics/Revue canadienne d'agroeconomie. 1987;**35**:591-603. DOI: 10.1111/j.1744-7976.1987.tb02251.x

[20] Shah AN, Tanveer M, Shahzad B, Yang G, Fahad S, Ali S, et al. Soil compaction effects on soil health and crop productivity: An overview. Environmental Science and Pollution Research. 2017;**24**:10056-10067. DOI: 10.1007/s11356-017-8421-y

[21] Ghosh U, Daigh AL. Soil compaction problems and subsoiling effects on potato crops: A review. Crop, Forage and Turfgrass Management. 2020;**6**:1-10. DOI: 10.1002/cft2.20030

[22] Mellgren PG. Terrain Classification for Canadian Forest, Canadian Pulp and Paper Association, Brossard, Quebec: OCLC 37823836; 1980. pp. 1-13

[23] Nawaz MF, Bourrie G, Trolard F. Soil compaction impact and modelling. A review. Agronomy for Sustainable Development. 2013;**33**:291-309. DOI: 10.1007/s13593-011-0071-8

[24] Smith CW, Johnston MA, Lorentz S. Assessing the compaction susceptibility of south African forestry soils. I. the effect of soil type, water content and applied pressure on uni-axial compaction. Soil and Tillage Research. 1997;**41**:53-73. DOI: 10.1016/S0167-1987(96)01084-7

[25] Jasa P, "Avoiding Harvest Compaction in Wet Soils," University 0f Nebraska-Lincoln Institue of Agriculture and Natural Resources, 2019. [Online]. Available from: https://cropwatch.unl.edu/2019/avoiding-compaction-harvest [Last Accessed 29 November 2021]

[26] Shawver C, Brummer J, Ippolito J, Ahola J and Rhoades R. "Managing Cattle Impacts When Grazing on Wet Soils." 2020. [Online]. Available from: https://extension.colostate.edu/topic-areas/agriculture/managing-cattle-impacts-when-grazing-on-wet-soils-1-634/ [Last Accessed 28 November 2021]

[27] National Inventors Hall of Fame. "Benjamin Holt Track-Type Tractor." 2006. [Online]. Available from: https://www.invent.org/inductees/benjamin-holt. [Last Accessed 30 November 2021]

[28] Caterpillar. "1900s," Peoria, IL: Caterpillar, Inc.; 2021. [Online]. Available from: https://www.caterpillar.com/en/company/history/1900.html. [Last Accessed 26 November 2021]

[29] Case IH. "Steiger & Quadtrac," CNH Industrial America, LLC. 2021. [Online]. Available from: https://www.caseih.com/emea/en-za/products/tractors/steiger-quadtrac-series/steiger-quadtrac [Last Accessed 23 November 2021]

[30] Lamandé M, Greve MH, Schjønning P. Risk assessment of soil compaction in Europe – Rubber tracks or wheels on machinery. Catena. 2018;**167**: 353-362. DOI: 10.1016/j.catena.2018.05.015

[31] Hawkins EM. Benchmarking costs of fixed-frame, articulated, and tracked tractors. Applied Engineering in Agriculture. 2015;**31**(5):741-745. DOI: 10.13031/aea.31.11074

[32] Lamandé M, Greve MH, Schjønning P. Risk assessment of soil compaction in Europe – Rubber tracks or wheels on machinery. Catena. 2018;**167**: 353-362. DOI: 10.1016/j.catena.2018.05.015

[33] Elmer's Manufacturing, Inc., "The benefits of tracks vs. tires," 2016. [Online]. Available from: https://elmersmfg.com/2016/04/benefits-tracks-vs-tires/. [Last Accessed 24 November 2021]

[34] NTS Tire Supply Team. "Tires vs. Tracks: Which Creates Less Compaction?" 2019. [Online]. Available from: https://www.ntstiresupply.com/ptk-shared/tires-vs-tracks-which-creates-less-compaction [Last Accessed 1 December 2021]

[35] Norman P. "IF and VF tyres - what are they?," Brocks Wheel & Tyre, 2021. [Online]. Available from: https://bwt.uk.com/news/if-tyres-and-vf-tyres-what-are-they/. [Last Accessed 21 November 2021]

[36] Tuschner J. "Compare & Contrast — Making the Case for Tires vs. Tracks," Farm Equipment, 2020. [Online]. Available from: https://www.farm-equipment.com/articles/18193 [Last Accessed 30 November 2021]

[37] Keller T, Arvidsson J. Technical solutions to reduce the risk of subsoil compaction: Effects of dual wheels, tandem wheels and Tyre inflation pressure on stress propagation in soil. Soil and Tillage Research. 2004;**79**: 191-205. DOI: 10.1016/j.still.2004.07.008

[38] Titan International, Inc., "LSW Technology," 2014. [Online]. Available from: https://www.titan-intl.com/innovation/lsw-tires. [Last Accessed 2 December 2021]

[39] Lyasko M. Slip sinkage effect in soil–vehicle mechanics. Journal of Terramechanics. 2010;**47**:21-31. DOI: 10.1016/j.jterra.2009 .08.005

[40] Abu-Hamdeh NH, Carpenter TG, Wood RK, Holmes RG. Soil compaction of four-wheel drive and tracked tractors under various draft loads. In: International Off-Highway & Powerplant Congress & Exposition - Milwaukee, Warrendale, PA: Society of Automotive Engineers. 1995. DOI: 10.4271/952098

[41] Moinfar A, Shahgholi G, Abbaspour-Gilandeh Y, Herrera-Miranda I, Hernandez-Hernandez JL, Herrera-Miranda MA. Investigating the effect of the tractor drive system type on soil behavior under tractor tires. Agronomy. 2021;**11**(4):696. DOI: 10.3390/agronomy11040696

[42] Davies DB, Finney JB, Richardson SJ. Relative effects of tractor weight and wheel-slip In causing soil compaction. Journal of Soil Science. 1973;**24**:399-409. DOI: 10.1111/j.1365-2389.1973. tb00775.x

[43] Vantsevich VV, Blundell MV. In: Vantsevich VV, Blundell MV, editors. Advanced Autonomous Vehicle Design

for Severe Environments. Amsterdam: IOS Press; 2015

[44] Kassel S. Lunokhod-1 Soviet Lunar Surface. Santa Monica: RAND Corporation; 1971. Available from: https://www.rand.org/pubs/reports/R0802.html

[45] Wikipedia. "Perseverance (rover)." The Wikipedia Project, 2021. [Online]. Available from: https://en.wikipedia.org/wiki/Perseverance_(rover) [Last Accessed 26 November 2021]

[46] Hoseinian SH, Hemmat A, Esehaghbeygi A, Shahgoli G, Baghbanan A. Development of a dual sideway-share subsurface tillage implement: Part 2. Effect of tool geometry on tillage forces and soil disturbance characteristics. Soil and Tillage Research. 2021;**215**:105200. DOI: 10.1016/j.still.2021.105200

[47] Spoor G. Alleviation of soil compaction: Requirements, equipment and techniques. Soil Use and Management. 2006;**22**:113-122. DOI: 10.1111/j.1475-2743.2006.00015.x

[48] Adamchuk VI, Skotnikov AV, Speichinger JD, Kocher MF. Development of an instrumented deep-tillage implement for sensing of sil mechanical resistance. Transactions of the ASAE. 2004;**47**(6):1913-1919. DOI: 10.13031/2013.17798

[49] Winsor S. "Healthy Soil and Profits from Low-Till." Farm Progress. 2012. [Online]. Available from: https://www.farmprogress.com/tillage/healthy-soil-and-profits-low-till [Last Accessed 23 November 2021]

[50] Alban DH, Host GE, Elioff JD, Shadis DA. Soil and Vegetation Response to Soil Compaction and Forest Floor Removal after Aspen Harvesting. Washington DC: USDA Forest Service; 1994. DOI: 10.2737/NC-RP-315

[51] Misser Uitgeverij BV. "NEXAT redesigning versatility, reducing field traffic," 2021. [Online]. Available from: https://www.futurefarming.com/tech-in-focus/autonomous-semiauto-steering/autonomous-vehicles/nexat-redesigning-versatility-reducing-field-traffic/ [Last Accessed 6 January 2022]

[52] Raper RL, Reeves DW, Burmester CH, Schwab EB. Tillage depth, tillage timing, and cover crop effects on cotton yield, soil strength, and tillage energy requirements. Applied Engineering in Agriculture. 2000;**16**(4):379-385. DOI: 10.13031/2013.5363

[53] Munkholm LJ, Schjonning P. Structural vulnerability of a sandy loam exposed to intensive tillage and traffic in wet conditions. Soil and Tillage Research. 2004;**79**:79-85. DOI: 10.1016/j.still.2004.03.012

[54] Dozier IA, Behnke GD, Davis AS, Nafziger ED, Villamil MB. Tillage and cover cropping effects on soil properties and crop production in Illinois. Agronomy Journal. 2017;**109**:1261-1270. DOI: 10.2134/agronj2016.10.0613

[55] Calonego JC, Raphael JP, Rigon JP, de Olieria-Neto L, Rosolem CA. Soil compaction management and soybean yields with cover crops under no-till and occasional chiseling. European Journal of Agronomy. 2017;**85**:31-37. DOI: 10.1016/j.eja.2017.02.001

[56] Williams SM, Weil RR. Crop cover root channels may alleviate soil compaction effects on soybean crop. Soil Science Society of America Journal. 2004;**68**:1403-1409. DOI: 10.2136/sssaj2004.1403

Chapter 7

Potential Applications of Rhizobacteria as Eco-Friendly Biological Control, Plant Growth Promotion and Soil Metal Bioremediation

Nafeesa Farooq Khan, Aatifa Rasool, Sheikh Mansoor, Sana Saleem, Tawseef Rehman Baba, Sheikh Maurifatul Haq, Sheikh Aafreen Rehman, Charles Oluwaseun Adetunji and Simona Mariana Popescu

Abstract

Modern agriculture has an immense problem in the depletion of agricultural productivity owing to a variety of biotic and abiotic stresses. Agriculture's sustainability and safety are dependent on ecologically friendly practices. Plant rhizobia have been proven to have an important role in disease control, as well as promoting plant growth, productivity, and biomass. Rhizobacteria are soil bacteria that live on the root surface and either directly or indirectly contribute to plant development. Rhizobia are used to induce mediated immune resistance through the manufacture of lytic enzymes, antibiotics, phytoalexins, phytohormone, metabolites. It supports the growth of plants through nitrogen fixation, nutrient enrichment, phosphate solubilization and phytohormone synthesis. In addition, it supports plants during different stresses such as temperature, osmotic, heavy metal and oxidative stress. Plant growth-promoting rhizobacteria have the ability to control heavy metal pollution of soils as well as enhancing plant growth in these soils. Efficient bioremediation is possible by using rhizobacterial inoculants, still, the distribution and functioning of microbes in the rhizosphere need to be fully explored. This review focuses on the effectiveness, biomonitoring processes and function in promoting plant development. Rhizobia application can be considered an alternative method for the improvement of biodiversity, agriculture, and the environment.

Keywords: rhizobia, biocontrol, antibiotic, plant growth promotion, heavy metal, bioremediation

1. Introduction

The productivity of crops is considerably impacted by nitrogen and phosphorous deficiencies, which are important for regulating the growth and development of crop plants [1]. To address this problem, it is important to carry out effective nitrogen management for sustainable agriculture. One of the interesting methods is to involve the use of microorganisms biologically fixing nitrogen which is utilized by the plant directly and is least susceptible to leaching and volatilization. Legumes establish a symbiotic interaction with the soil bacteria, termed Rhizobia, to fix atmospheric nitrogen. This helps in improving soil fertility, improving plant growth and prevents the necessity to use chemical fertilizers [2]. Besides this, agricultural productivity is significantly affected by the changing physical and biological properties of the soil [3]. In the past few years, the word "plant microsymbionts" has gained significant interest as plant microsymbionts directly affect the plant's performance and productivity. The plant microbiome comprises the complex adaptive gene pool, which originates from prokaryotic and eukaryotic organisms and even viruses, associated with the host's ecosystem [4]. Also, it has been well established that apart from changes in morphology, Bacteroides exhibit tremendous transcriptomic shifts and changes in biochemical processes especially in contrast to free-living bacteria [5]. There are various genetic and molecular pathways that govern the symbiotic compatibility, involving a wide variety of host and bacterial genes/signals with distinct adjuvants [6]. Consequently, understanding of the biological and molecular basis of symbiotic compatibility is essential in the development of tools for genetic modification of the host and/or bacteria to increase the efficiency of nitrogen fixation and to use it as a biocontrol agent. Here, in this review, we will address our latest summary of the microbial interactions, rhizobial efficacy, mechanisms as biocontrol, role in plant growth promotion, stress resistance and triggered immunity (ISR) against other microbes (pathogens). In fact, an insight into the genomes and recognition of candidate genes responsible for antibiotics, ISR and other metabolites from microbes is now possible. But the full range of molecular moieties involved in microbial interaction at an ecological scale deserves further study. Eventually, a definite and real improvement in the long term lies with the use of advanced analytical tools and their unification with classical experimental techniques to comprehend and then exploit soil–plant-microbe associations. Overall, it can help to improve biodiversity, agriculture and environmental studies further.

2. Microbial interactions

An existence of unseen host-microbial interaction has predominance from prehistoric times. While microbes are of minute size, they are available in nature in an astonishing majority, interacting directly or indirectly at different hierarchical levels of life. Almost all of these microorganisms are incredibly small, widely recognized by Archaea and Bacteria, although some microscopic forms include handful of fungi and even most protists. From an ecological standpoint, microorganisms are very often found in the soils as complex microbial population groups and have been investigated for several ties of microbiota-host interactions such as mutualists, endosymbionts, antagonists, parasites, and pathogens (**Figure 1**) [7].

Microbial community dynamic trends in the food chains look likely to be beneficial (positive), harmful (negative) or even sometimes neutral, with very little or no effect

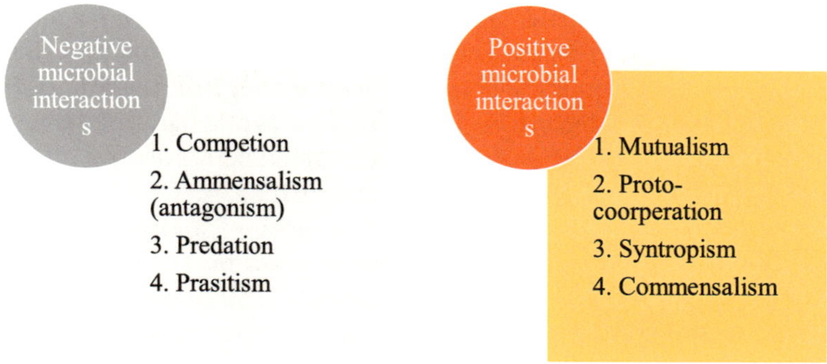

Figure 1.
Types of microbial interactions found in nature.

on their symbiotic associates [8]. Via physiochemical shifts, signaling mechanism-quorum sensing system (chemotaxis), cell transduction signaling through secondary metabolites, siderophores (used for iron acquisition) and gene expression microbial processes always have shown substantial impact on ecological parameters, resulting in established suitable alleles in diverse habitats [9]. Rapid and altered microorganism genetic variation corresponds both to biotic and abiotic sources of stress. Furthermore, atmospheric Nitrogen fixing microbial interaction and AMF symbiotic relationship activates a unique signaling process-CSSP (**C**ommon **S**ymbiosis **S**ignaling **P**athway) with calcium fluctuations in nucleus [10]. Many such strategies lead to an expansive population of microorganisms constantly getting established, culminating in pathogenic or beneficial effects on host plant species.

While many others have shown plants are able to select microbiota from all of diverse plant exudates including certain amino acids, carbohydrates and other biomolecules [11] which could also vary depending upon the plant itself, its stage of development and on biotic or abiotic conditions. Flavonoids, for example, are needed for talks between Legume-Rhizobia while AMF (mycorrhizal arbuscular fungi) rely solely on Strigolactone signaling [12]. In addition, the position of bacterial iron acquisition chelators that enforce a restricted supply of iron in the rhizospheric plane for pathogenic fungi constrains pathogen proliferation and occurrence. Consequently, synergetic microbial populations in the root micro-sites have a critical role to play in cloaking plants from disease deterioration, environmental factors and also ramping up nutrient uptake [13]. It has been well established that plant-associated microorganisms, particularly endophytic and rhizospheric microorganisms, can stimulate plant growth. A typical specified example is that of biotrophic symbioses between rhizobium and legume, such bacteria boost the growth of plant species by fastening atmospheric N_2, supplying of essential nutrients, enhance sequestration of minerals, produce phyto-hormones and also act as potential biocontrol against pathogens. Preliminary experiments on some endophytic and pathogen microbe genomes revealed pathogen degrade and displacement of host (host invasion), whereas the endophytic-mutualists express genes that aid in stress amelioration encoding proteins for nitrogen fixation and RubisCO [14]. During genetic interchange in a rhizobial symbiotic relationship, the root cortical cells are populated, making a distinction into nitrogen fixing bacteroids. Studies also show rhizobacter colonization into the root systems of non-leguminous plant species as such can be used as biocontrol in plant species other than legumes. Other popular, well-known, bacterial-based biocontrol method is *Agrobacterium* to prevent infection with *Agrobacterium tumefaciens.* In fact, myriad microorganisms (in particular belonging to genera *Bacillus, Pseudomonas* and *Trichoderma*) generate few chemicals against plant pathogenic fungi [8]. Bacterial isolates broadly find their application against plant pathogenic bacteria and fungi, whereas fungi are taken as biocontrols for pathogenic protozoans, pathogenic bacteria as well as pathogenic fungi. Juxtaposition between plants and several types of microbes has also been known to help mitigate many toxic metal build-up in plants [15]. While a general mechanism affecting mostly saprotrophs involves enhancement of microbial activity, selective different categories of symbionts can be stimulated in root microsites of plants. On the other hand, disease development by saproptrophs or biotrophs present in root micro-sites takes place only by developing antagonistic symbioses between pathogens and susceptible host plant roots. Importantly, the elimination of disease can sometimes be addressed through manipulating microbiological or physio-chemical surroundings mostly by classical practices- like use of soil refinements, agronomic rotational practices, fumigant use or even soil solarisation. A voluminous literature shows that interactive bacteria both symbiotic and pathogenic develop common signaling molecules to promote their host cell invasion through predominant substances such as conserved PAMP/MAMPs (Microbe-Associated Molecular Patterns) and protein effectors [16]. Organisms have developed recognizing mechanisms which differentiate between pathogens and symbionts and react in different ways to them, but this distinction often is not efficient; as a consequence, recognizing sensitivity also appears to occur both on pathogenic and symbiotic interaction [17, 18] at earlier stages. Thus, evidently microbial associations drive a complex sequence of interdependent metabolisms. In this paradigm of unexpected symbiotic partnership only host species utilize chemical synthesis capacities of symbiotic organisms to inhibit the development of certain environmental major competitors in order to sustain themselves [19]. In modern days, the philosophy of regulation of soil-borne diseases through the use of agro-chemicals such as pesticides and fungicides is now being modified through biological management [20]. Currently with the aid of molecular know-how, molecular pathways and processes involved in the interaction of microbes have been immensely explored.

3. Pathogen control mechanisms

Phytopathogens are those organisms which have the potential to adversely affect growth, development as well as the physiological activities of the crop. Any deviation in the environment which favors the proliferation of these phytopathogens result in a rapid outbreak of the diseases, leading to the crop destruction. Thus, reducing the yield and causing considerable loss of productivity. To prevent the development of disease it is necessary to control the pathogen mostly when their level is low. The organisms involved in biocontrol process are called as biocontrol agents and most of the biocontrol agents such as bacteria, fungi, algae, and nematodes which are found in root zone i.e., rhizosphere could influence various properties of soil and plants and thus act as defense mechanism against attack by pathogens [21]. It has been reported that there are some beneficial bacteria which can bring some changes in the rhizosphere as well as in the plants, leading to the enhancement in the plant growth, development and productivity and as such protect the plant from outbreak of various diseases [22]. Rhizobium being one of the categories of microorganisms which comprises of bacteria which can develop the symbiotic relationship with leguminous plants. Thus, are regarded as important nitrogen fixing organisms which play significant role in the maintenance of soil fertility [23]. However, many species of rhizobia are also reported to reduce the development of various disease-causing fungi, thereby increasing the yield of legume crops [24]. Several rhizobial strains such as *Rhizobium leguminosarum*, *Sinorhizobiummeliloti* and *Bradyrhizobium japonicum* have the ability to suppress soil-borne pathogens such as *Rhizoctonia solani, Pythium spp.*, *Fusarium spp.,* and *Macrophominaphaseolina* in both legumes and non-legumes thus can be used as biocontrol against various soil-borne diseases [25]. Godebo et al. [26] suggested that *rhizobium* species can be used as biocontrol agents, since it inhibited the growth of aphanomyces in vitro in pea. *Rhizobia in combination with Tricoderma spp. can act as potential biocontrol agent* [27]. Colonization behavior of *Sinorhizobium meliloti* in the alfalfa rhizosphere reported to be useful for biocontrol. The application of *Pseudomonas maltophilia* in combination with *Mesorhizobium* and PSB was reported to be more beneficial as it showed the reduction in root rot incidence [28]. The study above shows that significant reports have been presented which favors the use of rhizobia as biocontrol agent against soil-borne pathogens, apart from being responsible for biological nitrogen fixation, thus acting as a befitting alternate measure over chemical treatments to control the spread of various plant diseases. Rhizobium is an effective biocontrol agent which helps in bringing down the growth of phytopathogens by implementing various mechanisms which include phytohormone production, siderophore production, production of antibiotics, HCN production, production of lytic enzymes, metabolite production and phytoalexin production and induction of systemic resistance [29].

3.1 Lytic enzyme production

Rhizobia produces several lytic enzymes which are responsible for degrading the cell wall of pathogens and as such are considered as an efficient source for biocontrol. Lytic enzymes produced by the rhizobia for biocontrol involves chitinases, cellulases, β-1,3-glucanase β-1,4-glucanase, β-1,6-glucanase, proteases, pectinase and amylases [30]. These enzymes are known to cause lysis of the fungal and bacterial cell walls and thus helps in controlling the population of plant pathogens [31]. Chitinase is a lytic enzyme which causes the lysis of pathogenic fungal cell wall through the disintegration of chitin in the cell wall of fungi and bacteria. This process involves the breakdown of glycosidic bond in chitin thus, reducing the chitin polymer into monomer. Endochitinase cleaves chitin randomly at internal points within the polymer of chitin and releases low molecular weight multimers and dimers. Exochitinase causes hydrolysis of chitin and releases di-acetylchitobiose with no monosaccharide or oligosaccharides formed. Protease is another lytic enzyme which prevents the protein of pathogen to effect plant cells as protease have the capacity to cause the breakdown of proteins of phytopathogens into smaller polypeptides or single amino acids. Some of the protease also involved in inactivation of extracellular enzymes of phytopathogenic fungi. Cellulases is another enzyme which causes the decomposition of cellulose. This reaction involves the hydrolysis of the 1, 4-β-D-glucosidic linkages in cellulose. The degradation of cellulose involves conversion of the cellulose into β-glucose which occurs by the combined action of important cellulolytic enzymes like cellulose / endoglucanases, exo-cellobiohydrolase/exo-glucanases and β-glucosidases. Cellulose is thereby converted into β-glucose by the synergetic act of all these cellulolytic enzymes. Glucanase are enzymes which causes hydrolysis of polysaccharide made of glucose subunits. This process involves two possible mechanisms viz., cleaving the glucose residues from the non-reducing end in sequence and breaking the linkages along the polysaccharide chain at random points and smaller oligosaccharides are released [32]. Among all these enzymes, chitinase are considered to be the most important ones as it acts as prime constituent of biocontrol and protect the plant against phytopathogens. It has been reported that rhizobial isolates producing chitinase results in inhibition of pathogenic microbes [24]. Damping-off of fava bean (*Vicia faba*) was reduced when rhizobium spp. capable of producing chitinase was applied as seed treatment either separately or along with mycorrhizal fungi [24]. Rhizobium strains isolated from *Sesbania sesban* has been reported to be produce chitinase. Rhizobium sp. Strain RS12, which have the ability to produce chitinase controlled the diseases of chickpea caused by *F. oxysporum*, *S. sclerotiorum* and *M. phaseolina* by reducing the growth and development of mycelia [33]. Plant diseases caused by several phytopathogens like *A. niger, F. solani, F. oxysporium, B. cinerea* and *R. solani* were reported to be controlled by chitinase from rhizobia, thus the latter was regarded as efficient biocontrol agent (34). Ability of rhizobia to produce lytic enzymes such as chitinase, β-1, 3 glucanase, protease, and lipase which bring about the lysis of pathogenic fungal and bacterial cell walls was also reported in various plants [3]. In fava (*V. faba*) bean infection caused by fungal mycelia of *F. solani* was reduced significantly by chitinase, protease and lipase [34].

3.1.1 Phytohormone production

Phytohormones or plant hormones are the organic compounds that cause the stimulation of plant growth and development at lower concentrations. They can be produced either naturally by plants in response to some specific stimuli or can be synthesized artificially and utilized for regulating the growth and development of plants [35]. Apart from regulating growth and development, these phytohormones also play an important role in biocontrol responses as they are involved in several synergetic processes between various plants and organisms. Therefore, these plant hormones not only helps in stimulation of plant growth, development, improvement in nutrient uptake, but also act as a shield against various biotic and abiotic stresses, and as such protection of plants from different phytopathogens [36]. Phytohormones include indole-3-acetic (IAA) acid (auxin), cytokinins, gibberellins and abscisic acid. Each of the plant hormones or plant growth regulators possesses specific functions.

a. **Auxin:** This is the phytohormone which is considered as an important hormone that helps in plant protection mostly in the form of indole acetic acid (IAA). It has been suggested that many rhizobia spp. can secrete plant hormones, such as auxin via indole acetic acid formation [37–41]. Tryptophan has been considered as the major precursor of IAA. However, rhizobium spp. can synthesize IAA even if the tryptophan is not present [42]. Soil-beneficial bacteria have the ability to synthesize IAA and are involved in many phyto-stimulations that could be beneficial in relation to the biocontrol. IAA is also reported to loosen the root walls to increase the secretion of various beneficial substance from roots, which can improve the bacterial growth in root zone [22]. Rhizobia producing IAA are reported to directly affect the growth of phytopathogens (44). Rhizobial IAA is able to affect pathogenesis as being involved in various physiological processes of plant like cell division, extension, rate of xylem development, formation of adventitious root and various pigments, photosynthesis, etc. Therefore, can act as an effector molecule in plant microbial interaction. More than 80% of nitrogen-fixing bacteria have reportedly resulted in the production of growth substances like indole acetic acid [43]. These substances enhance plant defense mechanism against various pathogens and improves the plant growth by increasing the total phenols, calcium content and polyphenol oxidase activity [44]. Rhizobial IAA was reported to have Phyto stimulation activity which resulted in suppression of more than 84% fungus mycelial growth of *S. rolfsii* because of the synergetic relation between in vitro bacterial IAA production and inhibition of *S. rolfsii* mycelial [45]. Treatment of nodules of vetch roots with *R. leguminosarum* bv. *Viciae* resulted in increase of IAA production by about 60 folds [46]. Application of *Pseudomonas* in combination with *Rhizobium galegae* causes increase in IAA production that results in increasing the number of nodules, nitrogen content, growth of shoot and root. However, biosynthesis of IAA was influenced by both environmental stress factors (acidic pH, osmatic stress, matrix stress and carbon limitation) as well as by genetic factors (auxin biosynthesis genes and the mode of expression. The bacterial strain *Mesorhizobium loti* MP6 produces indole acetic acid (IAA) under normal growth conditions inducing curling of root hair, inhibition of *Sclerotiniasclerotiorum* and improves the growth of Indian mustard (*Brassica campestris*) [29].

b. **Gibberellins:** Gibberellins are plant hormones (GA1-GA89) that regulates various plant developmental processes having significant function in stem elongation and leaf expansion. Gibberellins are involved in many aspects of plant physiology like, development of seedless fruits, flower and fruit maturation, breaking of seed dormancy, and sex expression. It has been suggested that rhizobium also have ability to synthesis gibberellins. Gibberellic acid possesses the ability of reducing the levels of reactive oxygen species (ROS) which results in improving the activity of antioxidant enzyme which further causes the progress in growth under adverse conditions [47]. Also, gibberellic acid applied exogenously was able to reduce effect of various stress like salt, oxidative and heat stress, on growth and germination in *Arabidopsis thaliana*, resulting in increased production of salicylic acid, which in turn increased the activity of isochorismate synthase 1. Rhizobium strains are also reported to produce cytokinins, which are involved in stimulation of cell division, development of root and formation of root hair. It was established that microbial cytokinins have the potential to act as biocontrol agents and can be used as a potent source against plant defense mechanism [48].

c. ***Abscisic acid:*** Abscisic acid is a naturally occurring phytohormone. It is a sesquiterpenoid which is being partly produced in the chloroplasts of plants and the biosynthesis occurs in the leaves. Abscisic acid is synthesized mostly during the stress conditions like moisture deficiency and low temperatures, heat and salinity. It is reported that *rhizobium* sp. can produce abscisic acid and stimulate various physiological processes of plants such as stomatal closure, inhibits the shoot growth, storage of protein in seeds during dormancy and is involved in causing proteinase inhibition by gene transcription, thus offers protection against pathogens.

d. **1-aminocyclopropane-1-carboxylic acid (ACC) deaminase:** Some of the rhizobia species like α and β rhizobia have the ability to produce enzyme ACC deaminase and the gene responsible for its production is *acdS*gene. ACC deaminase leads to the conversion of 1-aminocyclopropane-1-carboxylic acid

(ACC-precursor of ethylene) into α-ketobutyrate and ammonia. It has been reported that when rhizobia producing ACC deaminase are inoculated, the ethylene levels in the plant are reduced, resulting in increased nodulation, longer roots as well as improves rhizobial activity and thereby helps in bringing down various stress levels and also protects the plant from various pathogens (**Table 1**). The strains, which are reported to produce ACC deaminase involve *R. leguminosarum. Viciae, Rhizobium hedysari, Rhizobium japonicum, Rhizobium gallicum, B. japonicum, Bradyrhizobium elkani, M. loti* and *S. meliloti* [59].

3.1.2 Antibiotics

Biologicals are an effective way of combating pathogens in plants [60]. Antibiotics and other antipathogenic compounds may be secreted by beneficial rhizobacteria. Antibiotics are among the most important pathways for biocontrol [61]. Pathogens also acquire antibiotic resistance and other biological control mechanisms to prevent complete long-term control. A systematic strategy of numerous monitoring mechanisms is definitely safer than undue reliance on one solution while confronting pathogens. Pathogen-antagonistic bacteria can therefore adapt their mode of

Rhizobium ssp.	**Activity**	**Reference**
Mesorhizobium cicero	IAA production	[49]
Rhizobium leguminosarum	IAA production	[50]
R. leguminosarum	Cytokinin	[51]
Mesorhizobium sp.	IAA production	[52]
Bradyrhizobium sp.	IAA production	[40]
Rhizobium sp.(lentil)	IAA production	[39]
Rhizobium phaseoli	IAA production	[53]
Bradyrhizobium sp.	IAA production	[43]
Rhizobium sp.	IAA production	[54]
Rhizobium sp. (pea)	IAA production	[55]
R. leguminosarum	IAA production	[48]
Mesorhizobiumloti MP6	IAA production	[29]
ACC deaminase		
R. japonicum, B. elkani, M. loti, R. leguminosarum, Sinorhizobium spp.	Produce high level of ACC deaminase	[3]
R. leguminosarumbv. Trifolii SN10	Produces indole acetic acid and ACC deaminase which enhances rice growth	[56]
Lytic acid production		
Rhizobium strain	Produce enzyme: chitinases, b-1,3 glucanases, proteases and lipases	[3, 57]
Rhizobium spp.	Chitinases	[24, 58]
Rhizobium sp. strain RS12	Chitinases	[33]

Table 1.
Phytohormone production.

operation in the long-term to combat pathogens. In order to inhibit pathogens, PGPR produces antibiotics, such as lipopeptides, polyketides, and antifungal metabolites [62]. PGPR generates antibiotics that prohibit "saprophytic pathogens" from developing in the root zone; Combining strains that strengthen resistance to other antibiotics and biocontrol strains that modulate one or more antibiotics [61]. Rhizobia produces (TFX) tridolitoxin, an antibiotic narrow-spectrum peptide, and was found responsible for changes in microbial diversity in bean plant rhizosphere. Trifolitoxin (TFX) antibiotic by *R. Leguminosarum*bv. *Trifolii* T24 was documented for disease control. *B. Japonicum* produces rhizobiotoxin which protects Soya from *M. Phaseolina* [63]. *R. Leguminosarum* produces bacteriocins which have different assumed size characteristics (small, medium or large). *Trifolii* and *B. Japonicum* secrete antibiotics that could inhibit several phytopathogens have been documented [3].

3.1.3 Phytoalexins

Plants exist in dynamic ecosystems which are subject to frequent changes. They survive on a host of chemicals called secondary metabolites [64], which are essential for regulating secondary metabolism. Plants have a normal immune system to withstand biotic stress which can be activated by different agents. The plants have a unique potential condition called "priming" which is triggered in the plant before the pathogen challenge. The plants defensive mechanism against biotic stress involves the agglomeration of molecules (phyto-anticipins), which are converted to phytoalexins [65]. Phytoalexins are antimicrobial compounds generated by plants or some organisms as a response of the biotic and abiotic factors. These are "low molecular weight, anti-microbial" compounds synthesized after micro-organism or abiotic exposure in plants. Furthermore, elucidating the biosynthesis of different phytoalexins allowed the use of molecular biology methods to investigate genes encoding enzymes involved in their synthesis. This has led to new technologies to improve plant resistance. Phytoalexins show enormous diversity in various chemical groups, such as terpenoids, phenolics, steroid glycoalkaloids, compounds containing sulfur and indoles [66].

3.1.4 Induced systemic resistance

In addition to its role in N fixation, rhizobium serves as a tool for biocontrol of plant pathogens by triggering systemic resistance in plants. This is referred to as Induced Systemic Resistance [67]. The latter prepares the plant for defense against various phytopathogens [68]. The mechanism by which a non-exposed part of a plant imparts resistance to pathogenic microbes etc. by earlier exposure with the former is termed as induced resistance, thus it is triggered by an inducer that can be a biological or chemical agent. This induced resistance is not only activated at the site of pathogen attack but also at the parts that are very far from the site of induction so called induced systemic resistance (ISR) (**Figure 2**) and this ISR provides resistance to broad spectrum pathogens. Systemic resistance provided by ISR is regulated by signaling pathways in which different hormones are involved [69].

Rhizobial species inducing systemic resistance are *Pseudomonas*, *Bacillus*, *Trichoderma* and *Mycorrhiza*. Stringlis et al. [70] observed that these rhizobia are involved in the biosynthesis of antibiotics, flagella, siderophores and other volatile compounds which in turn stimulate microbe associated molecular pattern triggered immunity (MTI). A signaling pathway is generated in response to the perception of any of the above-mentioned substances. This is followed by another signaling

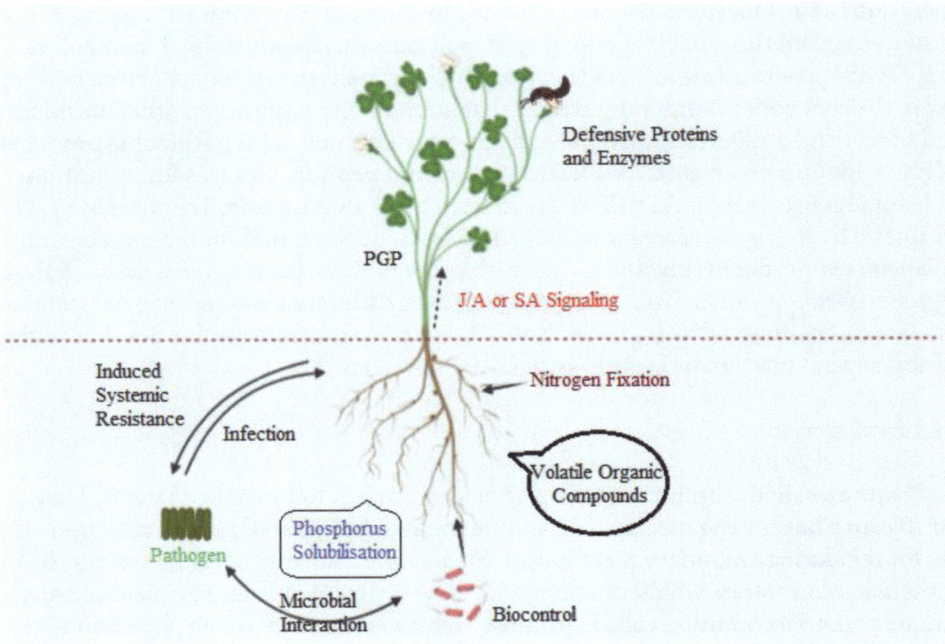

Figure 2.
Graphical representation of biologically induced disease resistance generated by beneficial microbes (ISR). It involves transport of long-distance signals in form of Jasmonic acid- salicylic acid (J/A & SA) and systemically circulate an improved defensive potential against a broad-spectrum pathogen in other plant parts and helps in plant growth promotion (PGP) as well.

pathway resulting in a systemic defense response [71]. Pattern-recognition receptors (PRRs) serve as sensors that have been evolved to differentiate and recognize bacterial and fungal products called pathogen associated molecular patterns (PAMPS). Moreover, in case of the damage/invasion caused by the pathogen attack an endogenous signal is produced. The ISR imitation in plants requires microbes that can be beneficial as well as able to effectively colonize the plants root system [72]. Recently microbial aspects around the root micro-sites harboring bacteria and fungi slowly gained interest because of their potential to trigger resistance (induced systemic resistance ISR in case of bacteria/systemic resistance in case of other microbes) in plants as a measure of biocontrol [17]. For instance, 22 kDa xylanase isolate of fungal endophyte *Trichoderma* when introduced into the plant cells evokes the plant's defensive response including potassium, hydrogen ions, calcium ion movements, PR protein synthesis, ethylene formation, glycosylation of phytosterols and fatty acid acylation [17]. Among the prominent changes taking place during ISR are:

1. Strength and stiffness in an epidermal and cortical plant cell wall.

2. Relocation of recently created barriers / blocks of impermeable lignin, callose and phenolic compounds away from an affected/entry site.

Plant responds to a number of biochemical signals induced by soil and plant-associated microbes. The strength and stability of its cross-talk signal play key role in determining the quality of resistance against pathogens. The interactions with these microbes can be in the form of different relationship possibilities (symbiosis, mutualism competition, predation, commensalism, etc. and host. At the initial stage, hypersensitive response gets active, a mechanism used by plants to prevent the spread of local infection by microbial pathogens [73]. While as for a positive mutual association both the host and the microbe must have to respond to the signals equally so that there is mutual benefit for both. In the association between the rhizobium and mycorrhiza, it has been studied that the host secretes strigolactones and flavonoids. Strigolactones are a class of plant hormones which are responsible for stimulation of branching and growth of mycorrhizal fungi. These strigolactones and flavonoids are also responsible for activation and production of symbiosis (sym) and Nodulaton (Nod) factors by microbes. The manipulated entry of rhizobium systematically triggers the whole downstream molecular defense system [67]. Which in turn builds a successful symbiotic relationship by activating common signaling pathways. By modifying the transcriptional programing many free-living plant growths promoting rhizobacteria (PGPR) positively respond to the root exudates that are involved in chemotaxis, energy metabolism etc. [74]. The mode of action of ISR is priming for enhanced defense, it does not cause direct activation of systemic resistance. Elevated transcript levels of various transcription factors were found in *Arabidopsis*eg. AP2/ERF were highly expressed. Among these several members are involved in regulation of jasmonic acid (JA) and ethylene (ET) defensive pathways. ISR by soilborne microbes is mostly regulated by JA/ET pathway. In the rhizosphere ISR is responsible for microbial antagonism, any host pathogen interaction enriches the microbiome and thus provides protection against diseases. The production of elicitors by beneficial microbes is also required in order to result in the onset of systemic immunity [69] so that there is a balance between the costs and benefits of mutualism. Plant-growth-promoting rhizobacteria (PGPR) were successful in managing complex diseases such as anthracnose (*Colletotrichum spp.*), angular leaf spot and bacterial wilt (*Erwinia tracheiphila*). Oxidative changes were observed in soyabean roots after inoculation with *Bradyrhizobium japonicum* [75]. With advancement of next generation sequencing technologies, it has been very easy to study the vast microbial diversity in the rhizosphere. Earlier studies have shown that there are different subsets of diversity in soil bulk, thus type of soil is an important factor for determining rhizosphere microbial community.

4. Mechanism in plant growth promotion

Modern agriculture is experiencing a number of challenges *viz*., poor soil fertility, serious pathogen and pest attacks, climate changes. Agricultural production must be sustainable and at the same time eco-friendly. This could be achieved by using environmentally sound approaches such as use of bio-fertilizers, bio-pesticides and by returning the crop residues to the soil thereby increasing the organic matter content of the soil. Application of crop residues to the soil resulted in increased yields compared to control [76]. Microbial inoculants which have been used for centuries, is a safer and relatively cheaper tool for promoting plant growth and improving soil health properties by different mechanisms [22]. Nitrogen fixing rhizobium bacteria live in association with legumes, infect them and form nodules in its roots. In case of non-legume crops they interact asymbiotically [77]. They are found in the rhizosphere to make use of the nutrients as the latter has plentiful nutrients oozed from roots of plants. They either have a direct or indirect control over plant growth, by synthesizing phytohormones, control pathogen infestation by influencing the production of several enzymes like cellulase, protease, lipase and other such productions thereby inducing whole plant resistance against pests or by soil nutrient enrichment through their nitrogen fixation and phosphate solubilizing ability. Microbial inoculants have multiple beneficial effects, particularly as plant growth promoters (PGP). Not only this but PGPR also help in combating a variety of abiotic stresses like temperature stress, salinity as well as drought stress, heavy metal toxicity and other types of abiotic stresses [3]. According to their closeness and interaction with the plant roots Rhizospheric bacteria have been classified as: (1) rhizosphere occupying bacteria (2) bacteria's forming colonies at the surface of roots (3) bacteria's living inside the roots (endophytes); and (4) bacteria's residing in the cells of root nodules. Bacteria's that belong to these groups are known as plant growth promoting rhizobacteria (PGPR) [78]. The bacteria belonging to 1 to 3 categories as extracellular PGPR (ePGPR) while the 4th category was named as intracellular PGPR (iPGPR). The ePGR includes following genera: *Bacillus*, *Pseudomonas*, *Erwinia*, *Caulobacter*, *Serratia*, *Arthrobacter*, *Micrococcus*, *Flavobacterium*, *Chromobacterium*, *Agrobacterium*, and *Hyphomicrobium* whereas *Rhizobium*, *Bradyrhizobium*, *Sinorhizobium*, *Azorhizobium*, *Mesorhizobium* and *Allorhizobium* belong to iPGR category. To strengthen the use of soil rhizobia for the attainment of sustainable and eco-friendly production methods a basic understanding of their functioning and means by which they facilitate plant growth is needed.

4.1 Plant growth promotion by direct mechanisms

4.1.1 Nutrient enrichment by Nitrogen fixation

Nitrogen is a macronutrient required by the plants for synthesizing proteins, nucleic acids and enzymes. Plants synthesize their food with the help of chlorophyll and nitrogen forms an essential component of chlorophyll. Despite the fact that the atmospheric air comprises of about 78% of nitrogen N, this gas is not available for use by the plants directly. Nitrogen application to crops has led to an enormous increase in food production which has eventually resulted in increased human population. Haber-Bosch process being the source of industrial nitrogen fertilizers, has been regarded as the primary cause of explosive growth in human population [79]. Currently, large amounts of synthetic chemical fertilizers are being used in agriculture and these fertilizers have been used beyond their limits, moreover they are expensive and polluting. Application of chemical fertilizers liberates reactive nitrogen into the atmosphere which leads to emission of green-house gases and at the same time eutrophication of water bodies. The detrimental effects of fertilizer use become much more pronounced when these are applied injudiciously. The economic and most importantly environmental concerns make the use safer and relatively cheaper alternatives necessary. Biological nitrogen fixation, whether symbiotic or non-symbiotic is a potential alternative promoting plant growth and hence increasing production [80]. Plant growth promoting-rhizobia are able to perform biological nitrogen fixation (BNF) and thus help plants in nitrogen assimilation. They live in soil and after producing specialized structures (nodules) in legumes by infecting their roots, they fix the atmospheric nitrogen (N_2) and convert the same into a more readily useable form i.e., ammonia (NH_3) so that the plants can utilize it for their growth. These rhizobia in turn get organic acids which serves as a source of carbon and energy. Two classes of genes: 1. Nodulation (nod) genes and 2. nitrogen fixation (nif) genes are needed

Function of the gene	Gene
Nodulation genes	
nodA	Acyltransferase
nodB	Chitooligosaccharide deacetylase
NodC	N-acetylglucosaminyltransferase
Nod	Transcriptional regulator of common nod genes
nodIJ	Nod factors transport
nodPQ	synthesis of Nod factors substituents
nodX	Synthesis of Nod factors substituents
nofEF	Synthesis of Nod factors substituents
Other nod genes	Several functions in synthesis of Nod factors
nol genes	Several Functions in synthesis of Nod factors substituents and secretion
NOE genes	Synthesis of Nod factors substituents
Nitrogen fixing genes	
nifHDK	Nitrogenase
NifA	Transcriptional regulator
nifBEN	Biosynthesis of the Fe-Mo cofactor
fixABCX	Electron transport chain to nitrogenase
fixNOPQ	Cytochrome oxidase
fixLJ	Transcriptional regulators
fixK	Transcriptional regulators
fixGHIS	Copper uptake and metabolism
fdxN	Ferredoxin

Table 2.
Genes involved in nitrogen fixation.

for the establishment of a good association between rhizobia and plants. Bacterial genes present in plasmids, code for Nod and Nif proteins [81]. Mainly three nod genes namely nodC, nodB and nodA are involved in nitrogen fixation. In addition to this, other nod genes viz., nod, nol or noe have been found in some rhizobial species [82]. Nodulation genes code for the enzymes involved in production of nodulation factors (nod) [77]. The roots of leguminous plants produce flavonoids in the rootzone, these compounds stimulate the expression of nod genes in the bacteria. Their expression in turn produces the Nod factor, which is a lipochito-ologosachharidic nodulation signal. This signal triggers mitosis and nodule formation [83]. Nitrogen fixation genes include genes for nitrogenase. Nitrogenase forms the most important part of BNF. The enzyme has 2 components: a. dinitrogenase reductase and b. dinitrogenase. The former gives electrons to the later which reduces N2 to NH3. BNF involves different clusters of genes for nitrogen fixation and nodule formation in leguminous plants (**Table 2**) [77].

4.1.2 Phosphate solubilization

Phosphorus is another macronutrient essential for proper development of plants. Its deficiency can adversely affect plant growth. After nitrogen phosphorous is the most limiting nutrient for plant growth [84]. Phosphorus forms an integral part of DNA and RNA, enzymes and phospholipids. Besides this, important processes like photosynthesis, formation of roots, flowers, ability of plants to cope up with diseases depend on the optimal levels of phosphorus [85, 86]. Although the soils are naturally rich in phosphorous reserves but the amount that is available to plants for their use is only a small fraction of the original amount present. This is because phosphorus is predominantly present in insoluble forms in soil and plants can only make use of phosphorus in soluble form i.e., the monobasic (H_2PO_4 -) and dibasic forms ($H_2PO_4{}^{2-}$). Phosphorus availability is governed by various factors such as pH of soil, soil temperature, amount of organic matter present in the soil, root system and most importantly soil microorganisms. The latter has a critical role in increasing P availability to plants. Soil P concentration ranges between 0.01-3 mg P $L^{-}1$ which is very small compared to the amount that plants need for normal growth. Therefore, to make sure that the plants are not devoid of P, remaining amount is compensated by soil rhizobia using their phosphate solubilizing property. These rhizobia are referred to as phosphate solubilizing microbes (PSMs), having the ability to hydrolyze insoluble phosphorus in soil into readily soluble form. They develop a network in the rhizosphere around the plant roots, allowing them to absorb P from a broader area. The use of PSMs is an environmentally safe and cheap method to reduce the insufficiency of phosphorous and promote its absorption and assimilation by plants. PSMs are able to convert the insoluble phosphorus into soluble form by lowering the pH, chelating cations and mineralization [84]. Application of phosphate solubilizing bacteria belonging to following genera: *Achromobacter, Agrobacterium, Bacillus, Pseudomonas, Erwinia, Flavobacterium, Microbacterium and Rhizobium* has resulted in increased phosphorus uptake and eventually higher yields.

4.1.3 Potassium solubilization

A diverse range of soil microorganisms such as saprophytic bacteria, fungi, and actinomycetes show potential to solubilize potassium effectively converting soil K to plant-available forms [87–90]. Among these, solubilizing bacteria (KSB) can dissolve K-rich materials and convert insoluble K to soluble forms that plants can absorb. Although some KSB can work anaerobically, the majority of these are aerobic. The potassium solubilizing rhizobacteria (KSR) use a number of ways to make the K available to plants. Mechanisms such as Acidolysis, chelation, exchange reactions, complexolysis, and the production of organic acids are few well known alternatives. The acidolysis (organic and inorganic acids, as well as the synthesis of protons) is the main mechanism of K mineral solubilization [87, 91–95]. Formation of organic acids by KSB that are useful in releasing K from K-bearing minerals include oxalic acid, tartaric acids, gluconic acid, 2-ketogluconic acid, citric acid, malic acid, succinic acid, lactic acid, propionic acid, glycolic acid, malonic acid and fumaric acid [96–103]. Tartaric acid, citric acid, succinic acid, ketogluconic acid, and oxalic acid are the most effective acids secreted by KSB among the several organic acids involved in the solubilization of insoluble K. *Acidothiobacillus ferrooxidans, Paenibacillus spp., Bacillus mucilaginosus, Bacillus edaphicus,* and *Bacillus circulans* are among the bacteria that can solubilize K minerals such as biotite, feldspar, illite, muscovite, orthoclase, and mica [96, 104]. It has been observed that *B. mucilaginosus, B. circulanscan, B. edaphicus, Burkholderia, A. ferrooxidans, Arthrobacter sp., Enterobacter hormaechei, Paenibacillus mucilaginosus, Paenibacillus frequentans, Cladosporium, Aminobacter, Sphingomonas, Burkholderia, and Paenibacillus glucanolyticus* solubilize K from silicate rocks. Further, *B. mucilaginosus, B. edaphicus,* and *B. circulanscan* have been identified as excellent K solubilizers in soil bacterial populations [88, 89]. Furthermore, microbial degradation of organic materials produces ammonia and hydrogen sulphide, both of which can be oxidized in the soil to make powerful acids like nitric acid (HNO_3) and sulfuric acid (H_2SO_4). Consequently, K+, Mg_2+, Ca_2+, and Mn_2+ are displaced from the cation-exchange complex in soil by hydrogen ions [105]. Organic acids produced by KSB can liberate K ions from the K mineral via complexing agent Si_4+, Al_3+, Fe_2+, and Ca_2+ ions (chelating) linked with K minerals, additional to decreasing soil pH [106, 107]. In addition, accumulation of diverse extracellular polymers (mainly proteins and polysaccharides) has also been linked to the release of K from K-bearing minerals [99, 103, 108]. Such substances act as adhesive structures to the surface of minerals or rocks. Fresh microbial EPS (exopolysaccharides) solution, for example, accelerates the dissolution rate of feldspars by forming complexes with framework ions in solution (Welch and Vandevivere 1994). Other PGPRs (for example, IAA-producing bacteria) may also play a role in delivering K to plants via boosting root exudates [109].

Under greenhouse and field circumstances, studies have demonstrated that inoculating seeds and seedlings of many plants with KSB improves germination percentage, seedling vigor, plant development, yield, and K uptake [87, 88, 110–115]. Several studies show that KSB inoculation improves the growth of a variety of crops [101, 103, 112, 116–125]. Overall, studies indicate application of KSB as bio-fertilizers for agriculture development can reduce the usage of agrochemicals while also promoting sustainable crop production

4.1.4 ACC deaminase production

The infection caused in the roots by rhizobium bacteria during nodule formation results in stress conditions. Consequently ethylene, a stress regulating hormone, inhibits the infection put forth by the bacteria, besides restricting nodulation and root growth [126]. Specific genes are involved in the interaction mechanisms of Rhizopheric bacteria with the plants by means of which they influence their growth. One of these genes encoding for the enzyme ACC deaminase, is involved in cleaving ACC, the precursor of ethylene biosynthesis produced by plants. ACC deaminase degrades ACC into ammonium and ketobutyrate and prevents ethylene biosynthesis [127]. Under limited ethylene concentration, rhizobial colonization of the roots is enhanced which result in the formation of a greater number of nodules on the host plant. Horizontal Gene transfer allows the spread of acdS within the species [128]. However, the genetic analysis carried out by Nascimento et al. [129] revealed that acdS are inherited vertically during evolution. Glick, [22], confirmed that the bacteria which produce IAA synthesize high level of ACC deaminase which inhibits ethylene biosynthesis and promote plant growth, root nodulation and increase uptake of minerals from the soil. Rhizobial strains including *R.leguminosarum, R. hedysari, R. gallicum, B. elkani and S. meliloti* have been reported to synthesize ACC deaminase [3].

4.2 Plant growth promotion by indirect promotions

4.2.1 Salt stress and osmotic stress

Plant growth improvement has been of great concern since the beginning of agriculture. There are various abiotic factors including temperature, pH, heavy metal toxicity, salt stress which obstruct plant growth and crop productivity [130]. Among them salinity stress is a real hazard for plant growth and production. Under saline conditions plants uptake high amounts of salt which interferes with their physiological and metabolic processes which hampers their growth and makes their survival difficult. Reclamation of saline soils by conventional methods i.e., adding soil amendments like gypsum, calcium etc. do not help to overcome salinity stress completely, moreover they adversely affect the ecosystem. Therefore, for the enhancement of plant growth and productivity, development of sustainable and safer methods is of utmost importance [131]. Large number of microbes belonging to different genera of salt tolerant plant growth promoting rhizobacteria (ST-PGPR), present in the soil are able to tolerate salinity stress as well as promote plant growth [132]. These rhizobacteria (ST-PGPR) include genera *Pseudomonas, Enterobacter, Agrobacterium, Streptomyces, Bacillus, Klebiella* and *Ochromobacter* [133, 134]. Salt-tolerant rhizobium isolated from legumes growing in sand dune sand tree legume [135] were able to tolerate upto 2.5–3% of NaCl concentration. In 2018, Zhang et al. [136] isolated 305 bacterial strains and found that 162 out of 305 could grow in NaCl concentration of 150 g/l. For boosting nitrogen fixation and productivity in high salt containing soils co-inoculation of legumes with salt tolerant rhizobial bacteria is a sustainable solution. Under non saline and saline condition silicon was found to enhance growth and nitrogen fixation in leguminous plants [137].

4.2.2 Temperature stress

Worldwide climate change had led to an increase in temperature, which adversely effects plant growth and development. Elevated temperatures result in decreased rate of photosynthesis, negatively influence plant water relations, flower and fruit development. Soil rhizobia indirectly help plants to combat heat stress. Most rhizobia prefer an optimum temperature range of 25–30°C for their growth, however, during their life cycle they experience a temperature out of this range. The growth promotion effect of different PGPR strains in plants was attributed to their nitrogen fixing ability but these effects were noticed prior to the beginning of nitrogen fixation [138]. This shows that the favorable effects of rhizobium in alleviating temperature stress does not depend on nitrogen status. It is due to stimulation of genes to express under high temperature stress conditions. The expression of these genes is regulated by heat stress transcription factors (Hsfs) [139]. HSPs are a family of proteins that are induced by a sudden temperature rise, they include chaperones and proteases, which confer high temperature tolerance to bacteria and thus contribute to the tolerance mechanism [140]. A microarray study conducted in *Sinorhizobiummeliloti* showed that 169 genes, which included the genes coding for HSPs and chaperones, were up regulated under high temperature conditions. Chaperones, like DnaK–DnaJ and GroEL–GroES, form an important component of the heat shock response. After heat shock, the hydrophobic domains of proteins are exposed, and they get denatured. These chaperons help the denatured proteins to get back to their original conformation [141]. The increased expression of chaperone genes was induced in heat tolerant strains compared to the strains of the same species that were sensitive to heat. Under high temperature stress HSPs increase the stability of cell membrane, thereby conferring heat tolerance to both, rhizobacteria as well as the plant under stress. Breeding of heat tolerant or development of transgenic heat tolerant cultivars is a laborious and less economic method. Hence, the application of rhizobacterial inoculants to plants under temperature stress should be preferred as it is relatively cheaper and less time consuming. Various physiological and biochemical changes in plants, are induced by low temperature resulting in poor plant growth and low crop survival rates [142].

Rigidification of membranes due to the decreased fluidity of cell membrane is one of these changes that plants experience when exposed to chilling stress [143]. Response to cold shock results in the synthesis of cold shock proteins (CSPs). Rhizobia strains isolated from the wild relative of chickpea at low temperatures (9–15°C), successfully nodulated chickpea, indicating that it could serve as a potential microbial inoculant under low temperature conditions to maintain the normal functioning of plants. Symbiotic association of rhizobium with alfalfa enhances its tolerance to low temperature by regulating important physiological and metabolic processes. The oxidative enzymes were more active in AN (active nodules) and IN (inactive nodules) groups, providing higher cold tolerance to these plants [144].

4.2.3 Oxidative stress

Plants, in response to various kinds of environmental stresses such as biotic and abiotic stress produce reactive oxygen species (ROS). Examples of ROS are singlet oxygen ($^1O^2$), superoxide anion (O^{2-}), hydrogen peroxide (H_2O_2) and hydroxyl radical (OH-). Accumulation of reactive oxygen species (ROS) as a result environmental stress is detrimental for plant growth as they modify the primary cell constituents like DNA, lipids, proteins etc. [145]. PGPR reduce the deleterious effects of ROS by producing antioxidant enzymes [146, 147] which include peroxidase (POD), superoxide dismutase (SOD), catalase (CAT), nitrate reductase (NR) and glutathione reductase (GR) and thus help in maintaining plant growth and crop productivity [148]. Based on the results of Shen et al. [149] it could be concluded that due to the activation of antioxidant machinery by the rhizobium inoculants, their use is the most effective way for enhancing plant growth and mitigating stress induced by ROS.

4.2.4 Metal stress

Heavy metals occur naturally in soils; however, their increased quantity is undesirable and has become a global concern over the time [150]. Anthropogenic activities like atmospheric pollution, industrial waste disposal, mining, and other practices predominantly contribute to heavy metal toxicity [151]. Heavy metal toxicity leads to inhibition of chlorophyll biosynthesis and proteins required for proper growth of plants and their normal functioning. Plant growth promoting rhizobacteria have the ability to control heavy metal pollution of soils as well as enhancing plant growth in these soils [152]. Bacteria's producing siderophores promote plant growth besides enhancing their nutrient uptake potential under heavy metal stress conditions. Rhizobacteria have been found to release metal-chelating substances (siderophores) in rhizosphere by means of which they affect the bioavailability of toxic heavy metals and their uptake by plants significantly. They transform these compounds into a less toxic form and promote their precipitation, absorption or adsorption. Plant associated rhizobia can be used for bioremediation, as they enhance the phytoextraction and phytostablization potential of plants [153]. By phytoextraction, plants carry the contaminants from the soil with the help of their roots and eventually collect these contaminants in the aboveground parts of the plant [154]. Phyto-stablization on the other hand, immobilizes the soil contaminants. The contaminants either get adsorbed on the root surface or absorbed by the roots or they are transformed into less soluble compounds. Phytoremediation has been accelerated by the application of rhizobacterial species such as *Bacillus*, *Pseudomonas, Azotobacter* [155]. Thus, efficient bioremediation is possible by using rhizobacterial inoculants, still distribution and functioning of microbes in rhizosphere needs to be fully explored.

5. Conclusion and future prospective

Rhizobia have enormous potential in terms of innovative and more sustainable crop management approaches; yet, we only comprehend a small portion of this potential. The effectiveness of strains of rhizobia documented in this chapter emphasizes the unique qualities of plant growth induction, defense pathways, and the resilience spectrum available against different environmental stresses on a wide range of agricultural crops. Although it is the most investigated bacteria which finds its application in agriculture practices but only few strains are widely known for their efficiency and effective application in disease management, nutrient uptake and signaling compounds they produce. These are often used for promoting plant development, particularly in challenging situations like heat and drought, which are becoming more common as climate change proceeds. The discovery of such possible rhizobia strains, as well as the development of a viable technology for use by agricultural producers, are still in their early stages. Thus, we conclude that a definite and real improvement in the long term lies with the use of advanced analytical tools and their unification with classical experimental techniques to comprehend and then further exploit soil–plant-microbe associations for ecofriendly and enhanced crop production. The identification of such promising rhizobia strains would allow for the extension of this study area, as well as improved agricultural sustainability.

Conflict of interest

The authors declare no conflict of interest.

Author details

Nafeesa Farooq Khan[1,†], Aatifa Rasool[2], Sheikh Mansoor[3], Sana Saleem[4], Tawseef Rehman Baba[2], Sheikh Maurifatul Haq[1], Sheikh Aafreen Rehman[5], Charles Oluwaseun Adetunji[6] and Simona Mariana Popescu[7]*

1 Department of Botany, University of Kashmir, Srinagar, India

2 Division of Fruit Science, SKUAST, Kashmir, India

3 Division of Biochemistry, FBSc SKUAST, Jammu, India

4 Division of Vegetable Science, SKUAST, Kashmir, India

5 Division of Entomology, SKUAST, Kashmir, India

6 Department of Microbiology, Edo University Iyamho, Edo State, Nigeria

7 Department of Biology and Environmental Engineering, University of Craiova, Craiova, Romania

*Address all correspondence to: popescu_simona83@yahoo.com

† All authors have equal contribution.

References

[1] Nemadodzi LE, Araya H, Nkomo M, Ngezimana W, Mudau NF. Nitrogen, phosphorus, and potassium effects on the physiology and biomass yield of baby spinach (*Spinacia oleracea* L.). Journal of Plant Nutrition. 2017;**40**(14):2033-2044

[2] Ouma EW, Asango AM, Maingi J, Njeru EM. Elucidating the potential of native rhizobial isolates to improve biological nitrogen fixation and growth of common bean and soybean in smallholder farming systems of Kenya. International Journal of Agronomy. 2016;**2016**:7. Article ID: 4569241. DOI: 10.1155/2016/4569241

[3] Gopalakrishnan S, Sathya A, Vijayabharathi R, Varshney RK, Gowda CLL, Krishnamurthy L. Plant growth promoting rhizobia: Challenges and opportunities. Biotech. 2015;**5**:355-377

[4] Backer R, Rokem JS, Ilangumaran G, Lamont J, Praslickova D, Ricci E, et al. Plant growth-promoting rhizobacteria: Context, mechanisms of action, and roadmap to commercialization of biostimulants for sustainable agriculture. Frontiers in Plant Science. 2018;**9**:1473

[5] Zook D. Symbiosis—Evolution's co-author. In: Gontier N, editor. Reticulate Evolution. Cham, Switzerland: Springer; 2015. pp. 41-80

[6] Gellings CW, Parmenter KE. Energy efficiency in fertilizer production and use in efficient use and conservation of energy. In: Gellings CW, editor. Encyclopedia of Life Support Systems. Oxford, United Kingdom: EOLSS Publishers Co. Ltd.; 2016. pp. 123-136

[7] Faust K, Raes J. Microbial interactions: From networks to models. Nature Reviews Microbiology. 2012;**10**:538-550

[8] Tshikantwa TS, Ullah MW, He F, Yang G. Current trends and potential applications of microbial interactions for human welfare. Frontiers in Microbiology. 2018;**9**:1156

[9] Braga RM, Dourado MN, Araújo WL. Microbial interactions: Ecology in a molecular perspective. Brazilian Journal of Microbiology. 2016;**47**:86-98

[10] He J, Zhang C, Dai H, Liu H, Zhang X, Yang J, et al. A LysM receptor heteromer mediates perception of arbuscular mycorrhizal symbiotic signal in rice. Molecular Plant. 2019;**12**(12):1561-1576

[11] Haldar S, Sengupta S. Plant-microbe cross-talk in the rhizosphere: Insight and biotechnological potential. The Open Microbiology Journal. 2015;**9**:1

[12] Akiyama K, Matsuzaki K, Hayashi H. Transesquiterpenes induce hyphal branching in arbuscular nycorrhizal fungi. Nature. 2005;**435**:824-827

[13] Frey-Klett P, Burlinson P, Deveau A, Barret M, Tarkka M, Sarniguet A. Bacterial-fungal interactions: Hyphens between agricultural, clinical, environmental, and food microbiologists. Microbiology and Molecular Biology Reviews. 2011;**75**(4):583-609

[14] Hardoim PR, Van Overbeek LS, Berg G, Pirttilä AM, Compant S, Campisano A, et al. The hidden world within plants: Ecological and evolutionary considerations for defining functioning of microbial endophytes. Microbiology and Molecular Biology Reviews. 2015;**79**(3):293-320

[15] Tiwari S, Lata C. Heavy metal stress, signaling, and tolerance due to plant-associated microbes: An overview. Frontiers in Plant Science. 2018;**9**:452

[16] Wang D, Yang S, Tang F, Zhu H. Symbiosis specificity in the legume–rhizobial mutualism. Cellular Microbiology. 2012;**14**(3):334-342

[17] Whipps JM. Microbial interactions and biocontrol in the rhizosphere. Journal of Experimental Botany. 2001;**52**:487-511

[18] Zipfel C, Oldroyd GE. Plant signalling in symbiosis and immunity. Nature. 2017;**543**:328-336

[19] Seth EC, Taga ME. Nutrient cross-feeding in the microbial world. Frontier in Microbiology. 2014;**5**:350

[20] Deepa N, Sreenivasa MY. Biocontrol strategies for effective management of phytopathogenic fungi associated with cereals. In: New and Future Developments in Microbial Biotechnology and Bioengineering. Elsevier; 1 Jan 2019. pp. 177-189

[21] Brevik E, Cerdà A, Mataix-Solera J, Pereg L, Quinton J, Six J, et al. The interdisciplinary nature of soil. Soil. 2015;**1**(1):117-129

[22] Glick BR. Plant growth-promoting bacteria: Mechanisms and applications. Scientifica. 2012;**2012**:15. Article ID: 963401. DOI: 10.6064/2012/963401

[23] Herridge DF, Peoples MB, Boddey RM. Global inputs of biological nitrogen fixation in agricultural systems. Plant Soil. 2008;**311**:1-18

[24] Mazen MM, Nadia H. El-Batanony, Abd El-Monium MM, Massoud ON. Cultural filtrate of Rhizobium spp.and arbuscular mycorrhiza are potential biological control agents against root rot fungal diseases of faba bean. Global Journal of Biotechnology and Biochemistry. 2008;**3**(1):32-41

[25] Noreen R, Shafique HA, Ali SA, Habiba SV, Ara J, Ehteshamul-Haque S. Role of mungbean root nodule associated fluorescent Pseudomonas and rhizobia in suppressing the root rotting fungi and root knot nematode affecting chickpea (*Cicer arietinum* L.). Pakistan Journal of Botany. 2016;**48**(5):2139-2145

[26] Godebo AT, Germida JJ, Walley FL. Isolation, identification, and assessment of soil bacteria as biocontrol agents of pea root rot caused by *Aphanomyces euteiches*. Canadian Journal of Soil Science. 2020;**100**(3):1-11

[27] Parveen G, Noreen R, Safique HA, Sultana V, Ehteshamul-Haque S, Athar M. Role of Rhizobia in suppressing the root diseases of soybean under soil amendment. Planta Daninha. 2019;**37**:0100-8358

[28] Pathak DV, Sharma MK, Sushil K, Naresh K, Sharma PK. Crop improvement and root rot suppression by seed bacterization in chickpea. Archives of Agronomy Soil Science. 2007;**53**(3):287-292

[29] Chandra S, Choure K, Dubey RC, Maheshwari DK. Rhizosphere competent Mesorhizobium loti MP6 induces root hair curling, inhibits Sclerotinia sclerotiorum and enhances growth of Indian mustard (*Brassica campestris*). Brazil Journal of Microbiology. 2007;**38**:124-130

[30] Gupta CP, Kumar B, Dubey RC, Maheshwari DK. Chitinase mediated destructive antagonistic potential of *Pseudomonas aeruginosa* GRC1 against Sclerotinia sclerotiorum causing charcoal rot of peanut. BioControl. 2006;**51**:821-835

[31] Mabood F, Zhou X, Smith DL. Microbial signaling and plant growth promotion. Candian Journal of Plant Science. 2014;**94**(6):1051-1063

[32] Jadhav HP, Sayyed RZ. Hydrolytic enzymes of rhizospheric microbes in crop protection. MOJ Cell Science and Report. 2016;**3**(5):135-136

[33] Smitha M, Singh R. Biocontrol of phytopathogenic fungi using mycolytic enzymes produced by rhizospheric bacteria of Cicer arietinum. Indian Journal of Agricultural Biochemistry. 2014;**27**(2):215-218

[34] Tamiru G, Muleta D. The Effect of Rhizobia Isolates against Black Root Rot Disease of Faba Bean (Viciafaba L) Caused by *Fusarium solani*. The Open Agriculture Journal. 2018;**12**:131-147

[35] Sajjad Y, Jaskani MJ, Asif M, Qasim M. Application of plant growth regulators in ornamental plants: A review. Pakistan Journal of Agricultural Science. 2017;**54**(2):327-333

[36] Cho ST, Chang HH, Egamberdieva D, Kamilova F, Lugtenberg B, Kuo CH. Genome analysis of Pseudomonas fluorescens PCL1751: A rhizobacterium that controls root diseases and alleviates salt stress for its plant host. PLOS ONE. 2015;**10**:e0140231

[37] Ahemad M, Khan MS. Plant-growth-promoting fungicide-tolerant Rhizobium improves growth and symbiotic characteristics of lentil (*Lens esculentus*) in fungicide-applied soil. Journal of Plant Growth Regulation. 2011;**30**(3):334-342

[38] Ahemad M, Khan MS. Response of greengram [Vigna radiata (L.) Wilczek] grown in herbicide-amended soil to quizalafop-p-ethyl and clodinafop tolerant plant growth promoting Bradyrhizobium sp. (vigna) MRM6. Journal of Agricultural Science and Technology. 2011;**13**(7):1209-1222

[39] Ahemad M, Khan MS. Insecticide-tolerant and plant-growth-promoting Rhizobium improves the growth of lentil (*Lens esculentus*) in insecticide-stressed soils. Pest Management Science. 2011;**67**(4):423-429

[40] Ahemad M, Khan MS. Toxicological assessment of selective pesticides towards plant growth promoting activities of phosphate solubilizing *Pseudomonas aeruginosa*. Acta Microbiologica et Immunologica Hungarica. 2011;**58**(3):169-187

[41] Ghosh PK, De TK, Maiti TK. Production and metabolism of indole acetic acid in root nodules and symbiont (*Rhizobium undicola*) isolated from root nodule of aquatic medicinal legume Neptunia oleracea Lour. Journal of Botany. 2015:1-11. Article ID: 575067

[42] Wani PA, Khan MS, Zaidi A. Effect of metal tolerant plant growth promoting Bradyrhizobium sp. (vigna) on growth, symbiosis, seed yield and metal uptake by greengramplants. Chemosphere. 2007;**70**(1):36-45

[43] Ahmad F, Ahmad I, Khan MS. Screening of free-living rhizospheric bacteria for their multiple plant growth promoting activities. Microbiological Research. 2008;**163**(2):173-181

[44] Chowdhury AK. Control of Sclerotium blight of groundnut by growth substances. Crop Research (Hisar). 2003;**25**:355-359

[45] Volpiano CG, Lisboa BB, São José JFB, de Oliveira AMR, Beneduzi A, Passaglia LMP, et al. Rhizobium strains in the biological control of the phytopathogenic fungi Sclerotium (Athelia) rolfsii on the common bean. Plant and Soil. 2018;**432**:229-243

[46] Camerini S, Senatore B, Lonardo E, Imperlini E, Bianco C, Moschetti G, et al. Introduction of a novel pathway for

IAA biosynthesis to rhizobia alters vetch root nodule development. Archives of Microbiology. 2008;**190**(1):67-77

[47] Manjili FA, Sedghi M, Pessarakli M. Effects of phytohormones on proline content and antioxidant enzymes of various wheat cultivars under salinity stress. Journal of Plant Nutrition. 2012;**35**(7):1098-1111

[48] Großkinsky DK, Tafner R, Moreno MV, Stenglein SA, García de Salamone IE, Nelson LM. Cytokinin production by Pseudomonas fluorescens G20-18 determines biocontrol activity against Pseudomonas syringae in *Arabidopsis*. Scientific Reports. 2016;**6**:23310

[49] Wani PA, Khan MS, Zaidi A. Synergistic effects of the inoculation with nitrogen fixing and phosphate solubilizing rhizobacteria on the performance of field grown chickpea. Journal of Plant Nutrition and Soil Science. 2007a;**170**(2):283-287

[50] Dey R, Pal KK, Bhatt DM. Growth promotion and yield enhancement of peanut (*Arachis hypogaea* L.) by application of plant growth-promoting rhizobacteria. Microbiological Research. 2004;**159**(4):371-394

[51] Noel TC, Sheng C, Yost CK. Rhizobium leguminosarum as a plant growth promoting rhizobacterium: Direct growth promotion of canola and lettuce. Canadian Journal of Microbiology. 1996;**42**(3):279-283

[52] Ahemad M, Khan MS. Ecological assessment of biotoxicity of pesticides towards plant growthpromoting activities of pea (*Pisum sativum*)-specific rhizobium sp. StrainMRP1. Emirates Journal of Food and Agriculture. 2012b;**24**(4):334-343

[53] Ahemad M, Khan MS. Effects of pesticides on plant growth promoting traits of Mesorhizobium strain MRC4. Journal of the Saudi Society of Agricultural Sciences. 2012d;**11**(1):63-71

[54] Volpiano CG, Lisboa BB, Granada CE, José JFBS, de Oliveira AMR, Beneduzi A. Rhizobia for biological control of plant diseases. In: Kumar V, Prasad R, Kumar M, Choudhary DK, editors. Microbiome in Plant Health and Disease: Challenges and Opportunities. Singapore: Springer Singapore; 2019. pp. 315-336

[55] Zahir ZA, Shah MK, Naveed M, Akhter MJ. Substrate-dependent auxin production by Rhizobium phaseoli improves the growth and yield of Vigna radiata L. under salt stress conditions. Journal of Microbiology and Biotechnology. 2010;**20**(9):1288-1294

[56] Philippe R, Dreyfus B, Singh A. Indole acetic acid and ACC deaminase-producing Rhizobium leguminosarumbv. trifolii SN10 promote rice growth, and in the process undergo colonization and chemotaxis. Biology and Fertility of Soils. 2012;**48**(2):173-182

[57] Kim YC, Jung H, Kim KY. An effective biocontrol bioformulation against Phytophthora blight of pepper using growth mixtures of combined chitinolytic bacteria under different field conditions. European Journal of Plant Pathology. 2008;**120**:373-382

[58] Edulamudi P, Masilamani AJA, Gopal Divi VRS, Konada VM. Inhibition of phytopathogenic fungi by chitinase producing rhizobium isolates obtained from root nodules of macrotylomauniflorum (LAM.) Verde. Bangladesh Journal of Botany. 2018;**47**(1):161-164

[59] Madhaiyan M, Poonguzhali S, Ryu JH, Sa TM. Regulation of ethylene

levels in canola (*Brassica campestris*) by 1-aminocyclopropane-1-carboxylate deaminase-containing Methylobacteriumfujisawaense. Planta. 2006;**224**(2):268-278

[60] Velivelli SL, De Vos P, Kromann P, Declerck S, Prestwich BD. Biological control agents: From field to market, problems, and challenges. Trends in Biotechnology. 2014;**32**:493-496

[61] Compant S, Duffy B, Nowak J, Clément C, Barka EA. Use of plant growth- promoting bacteria for biocontrol of plant diseases: Principles, mechanisms of action, and future prospects. Applied Environmental Microbiology. 2005;**71**:4951-4959

[62] Prashar P, Kapoor N, Sachdeva S. Biocontrol of plant pathogens using plant growth promoting bacteria. In: Lichtfouse E, editor. Sustainable Agriculture Reviews. Berlin: Springer; 2013. pp. 319-360

[63] Chakraborty T, Montenegro MA, Sanyal SC, et al. Cloning of enterotoxin gene from Aeromonas hydrophila provides conclusive evidence of production of a cytotonic enterotoxin. Infection and Immunity. 1984;**46**:435-441

[64] Van den Ende W, El-Esawe SK. Sucrose signaling pathways leading to fructan and anthocyanin accumulation: A dual function in abiotic and biotic stress responses? Environmental and Experimental Botany. 2014;**108**:4-13

[65] Boue SM, Cleveland TE, Carter-Wientjes C, Shih BY, Bhatnagar D, McLachlan JM, et al. Phytoalexin-enriched functional foods. Journal of Agricultural and Food Chemistry. 2009;**57**(7):2614-2622

[66] Jeandet P. Phytoalexins current progress and future prospects. Molecules. 2015;**20**:2770-2774

[67] Tonelli ML, Figueredo MS, Rodríguez J, Fabra A, Ibañez F. Induced systemic resistance -like responses elicited by rhizobia. Plant and Soil. 2020;**448**:1-14. DOI: 10.1007/s11104-020-04423-5

[68] Pieterse CMJ, Zamioudis C, Berendsen RL, Weller DM, VanWees SC, Bakker PA. Induced systemic resistance by beneficial microbes. Annual Review of Phytopathology. 2015;**52**:347-375

[69] Pieterse CM, Van der Does D, Zamioudis C, Leon-Reyes A, Van Wees SC. Hormonal modulation of plant immunity. Annual Review of Cell and Developmental Biology. 2012;**28**:489-521

[70] Stringlis IA, Proietti S, Hickman R, Van Verk MC, Zamioudis C, Pieterse CMJ. Root transcriptional dynamics induced by beneficial rhizobacteria and microbial immune elicitors reveal signatures of adaptation to mutualists. Plant Journal. 2018;**93**(1):166-180

[71] Saijo Y, Loo EPI, Yasuda S. Pattern recognition receptors and signaling in plant–microbe interactions. Plant Journal. 2018;**93**(4):592-613

[72] Lugtenberg B, Kamilova F. Plant-growth-promoting rhizobacteria. Annual Review of Microbiology. 2009;**63**:541-556

[73] Mauch-Mani B, Baccelli I, Luna E, Flors V. Defense priming: An adaptive part of induced resistance. Annual Review of Plant Biology. 2017;**68**:485-512

[74] Matilla MA, Espinosa-Urgel M, Rodríguez-Herva JJ, Ramos JL, Ramos-González MI. Genomic analysis reveals the major driving forces of bacterial life in the rhizosphere. Genome Biology. 2007;**8**(9):R179

[75] Fernandez-Göbel TF, Deanna R, Muñoz NB, Robert G, Asurmendi S,

Lascano R. Redox systemic signaling and induced tolerance responses during soybean–*Bradyrhizobium japonicum* interaction: Involvement of nod factor receptor and autoregulation of nodulation. Frontiers in Plant Science. 2019;**10**:141

[76] Lu X. A meta-analysis of the effects of crop residue return on crop yields and water use efficiency. PLoSONE. 2020;**15**(4):1-18

[77] Laranjo M, Alexandre A, Oliveira S. Legume growth promoting rhizobia: An overview on the Mesorhizobium genus. Microbiological Research. 2014;**169**(1):2-17

[78] Glick BR. The enhancement of plant growth by free living bacteria. Canadian Journal of Microbiology. 1995;**41**(2):109-117

[79] Erisman JW, Sutton MA, Galloway J, Klimont Z, Winiwarter W. How a century of ammonia synthesis changed the world. Nature Geoscience. 2008;**1**:636-639

[80] Beatty PH, Good AG. Future prospects for cereals that fix nitrogen. Science. 2011;**333**(6041):416-417

[81] Remigi P, Zhu J, Young JPW, Masson-Boivin C. Symbiosis within symbiosis: Evolving nitrogen-fixing legume symbionts. Trends in Microbiology. 2016;**24**(1):6375

[82] Andrews M, Andrews ME. Specificity in legume-rhizobia symbioses. International Journal of Molecular Sciences. 2017;**18**(4):705

[83] Oldroyd GE, Murray JD, Poole PS, Downie JA. The rules of engagement in the legume-rhizobial symbiosis. Annual review of genetics. 2011;**45**:119-144

[84] Kalayu G. Phosphate solubilizing microorganisms: Promising approach as biofertilizers. International Journal of Agronomy. 2019;**2019**:7. Article ID: 4917256. DOI: 10.1155/2019/4917256

[85] Sharma SB, Sayyed RZ, Trivedi MH, Gobi TA. Phosphate solubilizing microbes: Sustainable approach for managing phosphorus deficiency in agricultural soils. Springer Plus. 2013;**2**:587

[86] Kumar A, Kumar A, Patel H. Role of microbes in phosphorus availability and acquisition by plants. International Journal of Current Microbiology and Applied Sciences. 2018;**7**(5):1344-1347

[87] Meena VS, Maurya BR, Verma JP. Does a rhizospheric microorganism enhance K+ availability in agricultural soils? Microbiology Research. 2014;**169**:337-347

[88] Meena VS, Maurya BR, Bahadur I. Potassium solubilization by bacterial strain in waste mica. Bangladesh Journal of Botany. 2015a;**43**:235-237

[89] Meena VS, Maurya BR, Verma JP, Meena RS, editors. Potassium Solubilizing Microorganisms for Sustainable Agriculture. New Delhi: Springer; 1 Jan 2016

[90] Etesami H, Emami S, Alikhani HA. Potassium solubilizing bacteria (KSB): Mechanisms, promotion of plant growth, and future prospects: A review. Journal of Soil Science and Plant Nutrition. 2017;**17**(4):897-911

[91] Sheng XF, Xia JJ, Chen J. Mutagenesis of the *Bacillus edaphicus* strain NBT and its effect on growth of chili and cotton. Agricultural Science China. 2003;**2**:409-412

[92] Sheng XF, Zhao F, He LY, Qiu G, Chen L. Isolation and characterization of silicate mineral-solubilizing *Bacillus*

globisporus Q12 from the surfaces of weathered feldspar. Canadian Journal of Microbiology. 2008;**54**:1064-1068

[93] Uroz S, Calvaruso C, Turpault MP, Frey-Klett P. Mineral weathering by bacteria: Ecology, actors and mechanisms. Trends in Microbiology. 2009;**17**:378-387

[94] Parmar P, Sindhu SS. Potassium solubilization by rhizosphere bacteria: Influence of nutritional and environmental conditions. Journal of Microbiological Research. 2013;**3**:25-31

[95] Meena VS, Maurya BR, Verma JP, Aeron A, Kumar A, Kim K, et al. Potassium solubilizing rhizobacteria (KSR): Isolation, identification, and K-release dynamics from waste mica. Ecology Engineering. 2015b;**81**:340-347

[96] Hu X, Chen J, Guo J. Two phosphate- and potassium-solubilizing bacteria isolated from Tianmu Mountain, Zhejiang, China. World Journal of Microbiology and Biotechnology. 2006;**22**:983-990

[97] Krishnamurthy HA. Effect of pesticides on phosphate solubilizing microorganisms [M. Sc.(Agric.) thesis]. Dharwad: University of Agricultural Sciences; 1989

[98] Keshavarz Zarjani J, Aliasgharzad N, Oustan S, Emadi M, Ahmadi A. Isolation and characterization of potassium solubilizing bacteria in some Iranian soils. Archives of Agronomy and Soil Science. 2013;**59**:1713-1723

[99] Liu D, Lian B, Dong H. Isolation of *Paenibacillus* sp. and assessment of its potential for enhancing mineral weathering. Geomicrobiology Journal. 2012;**29**:413-421

[100] Prajapati K, Sharma M, Modi H. Isolation of two potassium solubilizing fungi from ceramic industry soils. Life Science Leaflets. 2012;**5**:71-75

[101] Prajapati K, Sharma MC, Modi HA. Growth promoting effect of potassium solubilizing microorganisms on okra (*Abelmoscus esculantus*). International Journal of Agricultural Science and Research. 2013;**1**:181-188

[102] Saiyad SA, Jhala YK, Vyas RV. Comparative efficiency of five potash and phosphate solubilizing bacteria and their key enzymes useful for enhancing and improvement of soil fertility. International Journal of Science and Research Publications. 2015;**5**:1-6

[103] Sheng XF, He LY. Solubilization of potassium-bearing minerals by a wild-type strain of *Bacillus edaphicus* and its mutants and increased potassium uptake by wheat. Canadian Journal of Microbiology. 2006;**52**:66-72

[104] Mo B, Lian B. Interactions between *Bacillus mucilaginosus* and silicate minerals (weathered adamellite and feldspar): Weathering rate, products, and reaction mechanisms. Chinese Journal of Geochemistry. 2011;**30**:187-192

[105] Huang Z, He L, Sheng X, He Z. Weathering of potash feldspar by *Bacillus* sp. L11. Wei sheng wu xue bao. Acta Microbiologica Sinica. 2013;**53**:1172-1178

[106] Römheld V, Kirkby EA. Research on potassium in agriculture: Needs and prospects. Plant Soil. 2010;**335**:155-180

[107] Štyriaková I, Štyriak I, Galko D, Hradil P, Bezdička P. The release of iron-bearing minerals and dissolution of feldspars by heterotrophic bacteria of *Bacillus* species. Ceramics Silikaty. 2003;**47**:20-26

[108] Shelobolina E, Xu H, Konishi H, Kukkadapu R, Wu T, Blöthe M, et al.

Microbial lithotrophic oxidation of structural Fe (II) in biotite. Applied Environmental Microbiology. 2012;**78**:5746-5752

[109] Etesami H, Alikhani HA, Hosseini HM. Indole-3-acetic acid and 1-aminocyclopropane-1-carboxylate deaminase: Bacterial traits required in rhizosphere, rhizoplane and/or endophytic competence by beneficial bacteria. In: Maheshwari DK, editor. Bacterial Metabolites in Sustainable Agroecosystem. Springer International. 2015. pp. 183-258. DOI: 10.1007/978-3-319-24654-3_8

[110] Anjanadevi IP, John NS, John KS, Jeeva ML, Misra RS. Rock inhabiting potassium solubilizing bacteria from Kerala, India: Characterization and possibility in chemical K fertilizer substitution. Journal of Basic Microbiology. 2016;**56**:67-77

[111] Awasthi R, Tewari R, Nayyar H. Synergy between plants and P-solubilizing microbes in soils: Effects on growth and physiology of crops. International Research Journal of Microbiology. 2011;**2**:484-503

[112] Lynn TM, Win HS, Kyaw EP, Latt ZK, Yu SS. Characterization of phosphate solubilizing and potassium decomposing strains and study on their effects on tomato cultivation. International Journal of Innovation and Applied Studies. 2013;**3**:959-966

[113] Subhashini DV, Kumar A. Phosphate solubilising *Streptomyces* spp obtained from the rhizosphere of Ceriops decandra of Corangi mangroves. The Indian Journal of Agricultural Sciences. 2014;**84**(5):12-16

[114] Zhang A-m, Zhao G-y, Gao T-g, Wang W, Li J, Zhang S-f, et al. Solubilization of insoluble potassium and phosphate by *Paenibacillus kribensis* CX-7: A soil microorganism with biological control potential. Africa Journal of Microbiology Research. 2013;**7**:41-47

[115] Zhang C, Kong F. Isolation and identification of potassium-solubilizing bacteria from tobacco rhizospheric soil and their effect on tobacco plants. Applied Soil and Ecology. 2014;**82**:18-25

[116] Sheng X. Growth promotion and increased potassium uptake of cotton and rape by a potassium releasing strain of *Bacillus edaphicus*. Soil Biology and Biochemistry. 2005;**37**:1918-1922

[117] Han H-S, Lee KD. Effect of co-inoculation with phosphate and potassium solubilizing bacteria on mineral uptake and growth of pepper and cucumber. Plant Soil Environment. 2006;**52**:130

[118] Han HS, Lee KD. Phosphate and potassium solubilizing bacteria effect on mineral uptake, soil availability and growth of eggplant. Research Journal of Agriculture and Biology Science. 2005;**1**:176-180

[119] Sangeeth KP, Bhai RS, Srinivasan V. *Paenibacillus glucanolyticus*, a promising potassium solubilizing bacterium isolated from black pepper (*Piper nigrum* L.) rhizosphere. Journal of Spices and Aromatic Crops. 2012;**21**(2):118-124

[120] Youssef GH, Seddik WMA, Osman MA. Efficiency of natural minerals in presence of different nitrogen forms and potassium dissolving bacteria on peanut and sesame yields. Journal of American Science. 2010;**6**:647-660

[121] Abou-el-Seoud I, Abdel-Megeed A. Impact of rock materials and biofertilizations on P and K availability for maize (*Zea Maize*) under calcareous soil conditions. Saudi Journal of Biological Sciences. 2012;**19**:55-63

[122] Badr MA, Shafei AM, Sharaf El-Deen SH. The dissolution of K and P-bearing minerals by silicate dissolving bacteria and their effect on sorghum growth. Research Journal of Agriculture and Biological Sciences. 2006;**2**:5-11

[123] Basak B, Biswas D. Modification of waste mica for alternative source of potassium: Evaluation of potassium release in soil from waste mica treated with potassium solubilizing bacteria (KSB). LAP LAMBERT Academic Publishing. 2012

[124] Bagyalakshmi B, Ponmurugan P, Balamurugan A. Impact of different temperature, carbon and nitrogen sources on solubilization efficiency of native potassium solubilizing bacteria from tea (*Camellia sinensis*). Journal of Biological Research. 2012;**3**:36-42

[125] Abdel-Salam MA, Shams AS. Feldspar-K fertilization of potato (*Solanum tuberosum* L.) augmented by biofertilizer. Journal of Agricultural and Environmental Science. 2012;**12**:694-699

[126] Nukui N, Minamisawa K, Ayabe SI, Akoi T. Expression of the 1-aminocyclopropane-1-carboxylic acid deaminase gene requires symbiotic nitrogen-fixing regulator gene nifA2 in Mesorhizobium loti MAFF303099. Applied and Environmental Microbiology. 2006;**72**(7):4964-4969

[127] Checcucci A, Azzarello E, Bazzicalupo M, De Carlo A, Emiliani G, Mancuso S, et al. Role and regulation of ACC deaminase gene in Sinorhizobiummeliloti: Is it a symbiotic, rhizospheric or endophytic gene? Frontiers in Genetics. 2017;**8**:6

[128] Lemaire B, Van Cauwenberghe J, Chimphango S, Stirton C, Honnay O, Smets E, et al. Recombination and horizontal transfer of nodulation and ACC deaminase (acdS) genes within Alpha- and Betaproteobacteria nodulating legumes of the Cape Fynbos biome. FEMS Microbiology Ecology. 2015;**19**(11):118

[129] Nascimento FX, Rossi MJ, Soares CRFS, McConkey BJ, Glick BR. New insights into 1-Aminocyclopropane-1-Carboxylate (ACC) deaminase phylogeny, evolution and ecological significance. PLoS ONE. 2014;**9**(6):e99168

[130] Ahmad P. Oxidative Damage to Plants: Antioxidant Networks and Signaling. Cambridge, MA: Academic Press; 2014

[131] Egamberdieva D, Wirth S, Bellingrath-Kimura SD, Mishra J, Arora NK. Salt-tolerant plant growth promoting rhizobacteria for enhancing crop productivity of saline soils. Frontiers in Microbiology. 2019;**10**:2791

[132] Niu X, Song L, Xiao Y, Ge W. Drought-tolerant plant growth-promoting rhizobacteria associated with foxtail millet in a semi-arid agroecosystem and their potential in alleviating drought stress. Frontiers in Microbiology. 2018;**8**:2580

[133] Singh RP, Jha PN. The multifarious PGPR Serratia marcescens CDP-13 augments induced systemic resistance and enhanced salinity tolerance of wheat (*Triticum aestivum* L.). PLoS One. 2016;**11**(6):1-24

[134] Sarkar A, Ghosh PK, Pramanik K, Mitra S, Soren T, Pandey S, et al. A halotolerant Enterobacter sp. displaying ACC deaminase activity promotes rice seedling growth under salt stress. Microbiological Research. 2018;**169**(1):20-32

[135] Ali SF, Rawat LS, Meghvansi MK, Mahna SK. Selection of stress tolerant Rhizobial isolates of wild legumes

growing in dry regions of Rajasthan, India. Journal of Agricultural and Biological Science. 2009;**4**(1):13-18

[136] Zhang S, Fan C, Wang Y, Xia Y, Xiao W, Cui X. Salt-tolerant and plant-growth-promoting bacteria isolated from high-yield paddy soil. Canadian Journal of Microbiology. 2018;**64**(12):968-978

[137] Etesami H, Adl SM. Can interaction between silicon and non–rhizobial bacteria benefit in improving nodulation and nitrogen fixation in salinity–stressed legumes? A review. Rhizosphere. 2020;**15**:100229

[138] Govindasamy V, Murugeasn S, Kumar U. PGPR-biotechnology for management of abiotic and biotic stresses in crop plants. In: Maheshwari DK, editor. Potential Microorganisms for Sustainable Agriculture. New Delhi: IK International Publishing; 2008. pp. 26-47

[139] Baniwal SK, Bharti K, Chan KY, Fauth M, Ganguli A, Kotak S, et al. Heat stress response in plants: A complex game with chaperones and more than twenty heat stress transcription factors. Journal of Bioscience and Bioengineering. 2004;**29**(4):471-487

[140] Patil A, Kale A, Ajane G, Sheikh R, Patil S. Plant growth-promoting rhizobium: Mechanisms and biotechnological prospective. In: Hansen AP, Choudhary DK, Agrawal PK, Varma A, editors. Rhizobium Biology and Biotechnology. Cham: Springer; 2017. pp. 105-134

[141] Hartl FU, Hayer-Hartl M. Converging concepts of protein folding in vitro and in vivo. Nature Structural & Molecular Biology. 2009;**16**:574-581

[142] Zhou A, Sun H, Feng S, Zhou M, Gong S, Wang J, et al. A novel cold-regulated gene from Phlox subulata, PsCor413im1, enhances low temperature tolerance in Arabidopsis. Biochemical and Biophysical Research Communications. 2008;**495**(2):1688-1694

[143] Hu ZR, Fan JB, Xie Y, Amombo E, Liu A, Gitau MM. Comparative photosynthetic and metabolic analyses reveal mechanism of improved cold stress tolerance in bermudagrass by exogenous melatonin. Plant Physiology and Biochemistry. 2016;**100**:94-104

[144] Yu-Shi L, Jin-Cai G, Yang S, Yi-Xin Z, Tian-Ming H, Pei-Zhi Y. Effect of rhizobium symbiosis on low-temperature tolerance and antioxidant response in Alfalfa (*Medicago sativa* L.). Frontiers in Plant Science. 2019;**10**:538

[145] Moller IM, Jensen PE, Hansson A. Oxidative modifications to cellular components in plants. Annual Review of Plant Biology. 2007;**58**:459-481

[146] Hidri IR, Barea JM, Mahmoud MB, Azcon AR. Impact of microbial inoculation on biomass accumulation by Sulla carnosa provenances, and in regulating nutrition, physiological and antioxidant activities of this species under non-saline and saline conditions. Journal of Plant Physiology. 2016;**201**:28-41

[147] Islam F, Yasmeen T, Arif MS, Ali S, Ali B, Hameed S, et al. Plant growth promoting bacteria confer salt tolerance in Vigna radiata by upregulating antioxidant defense and biological soil fertility. Plant Growth Regulation. 2016;**80**:23-36

[148] Ansari FA, Ahmad I, Pichtel J. Growth stimulation and alleviation of salinity stress to wheat by the biofilm forming *Bacillus pumilus* strain FAB10. Applied Soil Ecology. 2019;**143**:45-54

[149] Shen G, Ju W, Liu Y, Guo X, Zhao W, Fang L. Impact of urea addition and

rhizobium inoculation on plant resistance in metal contaminated soil. International Journal of Environmental Research and Public Health. 2019;**16**(11):1955

[150] Chibuike GU, Obiora SC. Heavy metal polluted soils: Effect on plants and bioremediation methods. Applied and Environmental Soil Science. 2014;**1**:1-12

[151] Lebrazi S, Fikri-Benbrahim K. Rhizobium-legume symbioses: Heavy metal effects and principal approaches for bioremediation of contaminated soil. In: Meena R, Das A, Yadav G, Lal R, editors. Legumes for Soil Health and Sustainable Management. Singapore: Springer; 2018. pp. 205-233

[152] Ojuederie OB, Babalola OO. Microbial and plant-assisted bioremediation of heavy metal polluted environments: A review. International Journal of Environmental Research and Public Health. 2017;**14**(12):1504

[153] Kong Z, Glick BR. The role of plant growth-promoting bacteria in metal phytoremediation. Advances in Microbial Physiology. 2017;**71**:97-132

[154] Mahar A, Wang P, Ali A, Awasthi MK, Lahori AH, Wang Q, et al. Challenges and opportunities in the phytoremediation of heavy metals contaminated soils: A review. Ecotoxicology and Environmental Safety. 2016;**126**:111-121

[155] Ma Y, Prasad MNV, Rajkumar M, Freitas H. Plant growth promoting rhizobacteria and endophytes accelerate phytoremediation of metalliferous soils. Biotechnology Advances. 2011;**29**(2):248-258

Chapter 8

Heavy Metal Contamination in Vegetables and Their Toxic Effects on Human Health

Seema Manwani, Vanisree C.R., Vibha Jaiman, Kumud Kant Awasthi, Chandra Shekhar Yadav, Mahipal Singh Sankhla, Pritam P. Pandit and Garima Awasthi

Abstract

Vegetables are a prevalent nutrition for people all over the world because they are high in important nutrients, antioxidants, and metabolites that function as buffers for acidic compounds created during digestion. Vegetables, on the other hand, absorbed both vital and poisonous substances through the soil. Possible human health concerns, including as cancer and renal damage, have been linked to the consumption of heavy metal-contaminated vegetables (HMs). Heavy metals like Cr, Mn, Fe, Ni, Cu, Zn, Cd, Pb, and Hg were found in high concentrations in popular vegetables such as *Amaranthus tricolour L., Chenopodium album L., Spinacia oleracea, Coriandrum sativum, Solanum lycopersicum, and Solanum melongena.* The toxicity, fortification, health hazard, and heavy metals sources grown in soil are detailed in this review study.

Keywords: vegetables, heavy metals, toxic effects, human health, contamination

1. Introduction

Heavy metals, which are a major environmental problem, have a natural residency in the continental mantle. In general, a heavy metal is nothing but any chemical element which is metallic with a comparatively higher density that is poisonous above a tolerable range, such as mercury (Hg), cadmium (Cd), chromium (Cr), nickel (Ni), lead (Pb), and so on [1–3]. Contaminated heavy metal is a key cause of pollution and a possible increasing environmental and human health hazard all over the world, resulting in disorders in people and animals by consuming polluted vegetables. Heavy metals have damaged soil and water eco-systems worldwide. Heavy metals have been discharging into the environment through a variety of practises, including irrigation with polluted water, the use of chemical-based fertilisers, the dumping of industrial effluents into bodies of water, volcanic eruptions, forest fires, and so on [4]. Metals may seep into the ground, ground water, and eventually agricultural plants. Heavy metals can have serious consequences for human health when vegetables polluted with these metals are ingested. Although trace levels of copper (Cu), iron (Fe), manganese (Mn), nickel (Ni), and zinc (Zn) are needed in plants, excessive quantities of these metals can be hazardous [5, 6]. Metals including aluminium (Al), arsenic (As), cadmium (Cd), lead (Pb), and mercury (Hg) are not essential for regular human function and can cause toxicity promptly [7].

Vegetables are an integral portion of the normal diet because they contain nutritionally vital substances that are necessary for human existence. They also act as protective foods by contributing in the avoidance of disorders in people. Vegetables grown in areas polluted with dangerous metals or nearby sources of heavy metal pollution may gather greater amounts of heavy metals than other vegetables. Heavy metals are taken through the roots of plants from polluted soils and environmental wastes, entering the edible sections of plant tissues or accumulating on the surface of vegetables. Protracted irrigation of heavy metals with polluted garbage water raises heavy metal concentrations over the allowable limit [8].

The sensitivity, supplementation, potential dangers, and heavy metals sources grown in soil are all reviewed in this review study. Vegetables absorbed both essential and toxic chemicals from the soil. The consumption of heavy metal-contaminated vegetables has been related to potential human health issues such as cancer and kidney impairment (HMs). Heavy metals including Cr, Mn, Fe, Ni, Cu, Zn, Cd, Pb, and Hg were discovered in high amounts in common vegetables such *Amaranthus tricolour L., Chenopodium album L., Spinacia oleracea, Coriandrum sativum, Solanum lycopersicum, and Solanum melongena.*

2. Accumlation of heavy metals in vegetables

Mechanical, biochemical, and biological processes, as well as doings of human, could releases heavy metal into the environment and may cause heavy metal contaminants to accumulate inside living creatures in the food chain [9, 10]. HMs diffuse into the soil, air, as well as water bodies, wherever they could be had or eaten by crops/plants, bio-accumulating into upper consumers, and then biomagnified [11–13]. HMs cannot be easily removed from the top of the food chain once they have entered it, and they are thus cycled throughout the entire food web. Numerous hyperaccumulated plants provide nourishment for both humans as well as animals. As a result, the rotation from soil to humans thru plants and back into the soil following the expiry of upper consumers provides a pathway for HMs to persist in the environment for extended time periods, causing a variety of negative impacts. Ingestion vegetables containing HMs may provide potentially harmful health risks to lifeforms [14, 15] (**Figure 1**).

Heavy metals come in the food chain from a variety of sources. Cd, for example, took up from the soil by the roots and transferred to the body of plant. In the instance of Pb, the heavy metal is absorbing by plants through air pollution, whereas As and Hg can be received from dirt water. Some heavy metals having a capacity to accumulate in the tissues (liver, feathers, muscles, kidney, and other organs) of upper customers during the transit from one segment of the food chain to the next. Metal are liberated into the soil, water, and air from their parental material. These HMs are found in soil in decipherable, non-soluble, and moderately soluble forms, with the soluble forms being most harmful since they are quickly captivated by plants through roots before spreading all over the whole plant organs. Metal toxicity is caused through the disruption of cellular metabolic processes [16–19]. Hazardous metals are changed to persistent oxidation states in the acid standard and combine with particular proteins and enzymes when they reach the stomach from contaminated foods. The stabilised

Figure 1.
Vegetables get contaminated through various ways.

metal compounds interact with cysteine's sulphydryl groups (-SH) as well as methionine's sulphur atoms (-SCH3), causing protein molecules to breakdown [18, 20].

3. Impact of heavy metals on the quality of vegetables

Vegetation sensitivity to nutrition and metal concentrations varies, and their reactions can be seen in variations in stain concentration, liquid content, dehydrated weight, as well as development [21–23]. All of these variations in plant properties lead to different light absorption as well as reflectivity characteristics, which can be utilised to determine soil pollution and plant physiological condition. A few research findings have shown that metal and nutritional anxiety in plants contribute to differences in the supernatural reflectivity of the undergrowth [24–26], which may end up causing numerous biological effects in the plants and thus contribute to nutritional availability in veggies increasing or decreasing. Toxicity of metals in plants causes high germination inhibition, significant reductions in rates of growth, variations in photosynthetic efficiency, respiration, and transpiration, as well as changes in nutrient homeostasis and Mn, K, Mg. [27, 28] discovered distinctive leaf symptoms in Raphanus and Phaseolus, as well as a decrease in the root: shoot ratio and ratio of biomass. Higher levels of HM as well as cytochemical localization of Zn in *Raphanus* and *Pha-seolus*, which may cause stress, defence, and detoxification, are attributable to Zn's direct actions and the combined indirect effects of heavy metal (**Table 1**).

Sr. No.	Heavy metals	Vegetables	Observations	Area	References
1	Pb, Cd	*Spinacia oleracea and Solanum lycopersicum*	The concentration of the HMs increased than allowable limit	Amba nalla in Amravati city, Maharashtra	[29]
2	Pb, Cd, Cu, Zn, and As	*Raphanus sativus L., Daucus carota L., Ipomoea batatas L., Brassica parachinensis, Brassica campestris L., Solanum melongena L., Capsicum annuum Linn, Lycopersicum esculentum Mill, Momordica charantia Linn, Luffa cylindrical, Cucumis sativus, Cucurbita moschata Duch, Ipomoea aquatica Forsk, Amaranthus tricolour, Brassica oleracea, Brassica Chinensis Linn, Brassica pekinensis, S. oleracea, Coriandrum sativum, Lactuca satiua, Vigna unguiculata, and Phaseolus vulgaris*	Observing health problems	Shizhuyuan area in China	[30]
3	Cr, Ni, Cu, Zn, As, Cd, and Pb	*S. lycopersicum, Lagenaria siceraria, Solanummelongena, Cucurbita maxima, Amaranthus viridis L., Amaranthus paniculatus L., and Capsicumannuum L.*	Health risks of Cr, Cu, As, Cd, and Pb should be of great concern	Dhaka city, Bangladesh	[31]
4	As, Cd, Cr, Pb and Zn	*Lepidium sativum, Foeniculum vulgare, C. sativum, and Spinacea oleracea*	Pb and Cd levels exceeded the maximum permissible limits set by FAO/WHO for human consumption	Market sites of Kathmandu	[32]
5	Cd, Cu, Pb and Zn	*Lactuca satiua L., Spinacia Oleracea L., Allium ampeloprasum, Mentha, and Petroselinum crispum L.*	Cd and Pb levels exceeding the maximum level (ML) set by the Australian and New Zealand Food Authority	Port Kembla and Boolaroo, Australia	[33]
6	Fe, Zn, Cu, Pb, Cd, Mn, and Cr	*S. oleracea L., B. oleracea L. var. capitata Linn., B. oleracea L., S. melongena, Abelmoschus esculentus, Lycopersicum esculentum Mill, and R. sativus L.*	High level of pollution along cement factories of Rewa, India	J.P. Cement (Rewa)	[34]
7	Cu, Cd, Zn and Pb	*Beta vulgaris L., A. esculentus L. and B. oleracea L.*	The concentration of the HMs increased than allowable limit	Market sites of India	[35]

Table 1.
Heavy metals impacted vegetables from different areas.

Growth of plant was inhibited in both treatments of Cd, i.e. leaf chlorosis symptoms at 10 M Cd and necrotic patches at 100 M Cd, according to [36, 37], and browning of root was detected in both dealings. In root abstracts of Cd-exposed plants, the action of phosphoenolpyruvate carboxylase, which is involving in the anaplerotic fixation of CO2 into organic acids, increased. At 100 M Cd, citrate synthase, isocitrate dehydrogenase, and malate dehydrogenase activities increased significantly in leaf extracts, although fumarase activity declined. Membrane damage, electron transport disturbances, enzyme inhibition/activation, and interactions with nucleic acids are among known effects of metal toxicity [38, 39]. The production of oxidative stress and the substitution of critical cofactors of numerous enzymes, like Zn, Fe, and Mn, are two plausible causes for the development of these illnesses. Various researchers have associated oxidative stress with introduction to high heavy metal concentrations [40, 41]. Heavy metals' influence on plants, according to [42–44], growth suppression, physical harm, and a decay in physical, biological, and plant function are all consequences. Heavy metal toxicity disrupts cell and organelle membrane integrity by blocking enzymes, polynucleotides, and important nutrient and ion transport systems, displacing and/or substituting essential ions from cellular locations, denaturing and inactivating enzymes, and denaturing and inactivating enzymes. At supra-optimal absorptions, heavy metals as Cd, Pb, Hg, Cu, Zn, and Ni impede plant development, growth, and yield.

Interspecies distinctions in metal and nutrient uptake, as well as differences caused by therapeutic interventions within the similar plant, are minor and could be due to plant biomass and root exudes into the soils. The availability of metals and nutrients for plant absorption will be affected when plants develop and roots grow in soil due to biogeochemical interactions of organic acids generated by root oozes. This method may explain why tomato (*Solanum lycopersicum*) and pepper (*Piper nigrum*) plants absorb more Cu as well as Zn than other crops. According to [45, 46], Zn & Cu create organometallic compounds with organic acids found in root exudes, resulting in enhanced plant absorption. Excessive Zn in the growth media was shown to be hazardous to all 3 vegetable crops. Chlorosis in early leaves, searing of coralloid roots, and severe suppression of plant development were all signs of toxicity. With rising Zn concentrations, shoot fresh weight (FW) dropped.

Cu had a negative effect on seed germination in Chinese cabbage, according to [47]. (*Brassica pekinensis*). The germination rate was significantly lowered by the 0.5 mmol L1Cu treatment, with a median fatal dosage of 0.348 mmol L1. In early seedlings, Cu lowered root and shoot lengths, however the0.008 mmol L1treatment resulted in stimulatory elongation of the shoots. The aluminium coagulators had a toxic outcome on the plant growth of vegetable seeds at the tested concentrations. Furthermore, excessive copper levels in growing media harmed all 3 vegetable crops, causing chlorosis in new leaves, brown, stunted, coralloid roots, as well as plant development inhibition [48, 49].

Lin et al. [40] find that under higher Cd concentrations, the content of protein of desolate carrot (Daucus carota) and common sunflower (*Helianthus annuus*) decreased. Increased Zn concentrations reduced the content of protein of algae and Rapeseed (*Brassica napus*), according to [50]. The reduce in content of protein has been linked to increased protease activity speeding up protein breakdown [51, 52] or heavy metals interfering with nitrogen metabolism. Heavy metals, according to [53], may disrupt nitrogen metabolism, reducing protein synthesis in vegetables, and are also reason for a decrease in photosynthesis, which affects protein synthesis [40]. Cd could impair the absorption of Fe, potassium, Mn & calcium [54], and the toxicity amount had been observed to be higher in the case of specific heavy metals.

4. Intake of heavy metal in human body through vegetables

The industries are growing day by day in our country. The waste chemical contaminated water from these industries is directly thrown in river, sea, etc. Also the wastes, garbage from city is thrown in the water. This is the major reason behind the contamination of water. This water is used in many purposes like drinking, agriculture, etc. The contaminated water used in agriculture is absorbed by various vegetables. Resulting the vegetables become contaminated. We the humans use these contaminated vegetables in eating purpose. Once it is ingested in digestive system, it shows poisonous effects on the body as described in **Figure 2**. Heavy metal exposure typically follows this outline: from industries to air, soil, water, and foods, and then to people [55–59]. This heavy metals are existing in a amount of formats. Heavy metals like lead, cadmium, manganese, as well as arsenic could arrive the body by the gastrointestinal system or the entrance of digestive system while eating, drinking, or eating fruits and vegetables. The bulk of bodily heavy metals are transferred from blood to tissues [60, 61]. Red blood cells passes lead to not only the liver but also kidneys, where it is subsequently re-assigned as phosphate salt to the teeth, bone as well as hair [62–64]. Cadmium firstly fixes to blood cells & albumin, formerly to metallothionein in the kidney as well as liver. Later being carried through the blood to the lungs, vapour of manganese disperses over the membrane of lung to the central nervous system (CNS). Water solvable inorganic manganese ions are dispersed in the plasma as well as kidney for renal removal, whereas fat solvable manganese salts are diffused in the colon for faecal removal. Accumulation of Arsenic in the heart, lungs, liver, kidney, muscle, and neural tissues, as well as the skin, nails, and hair, afterward being passed by the circulation.

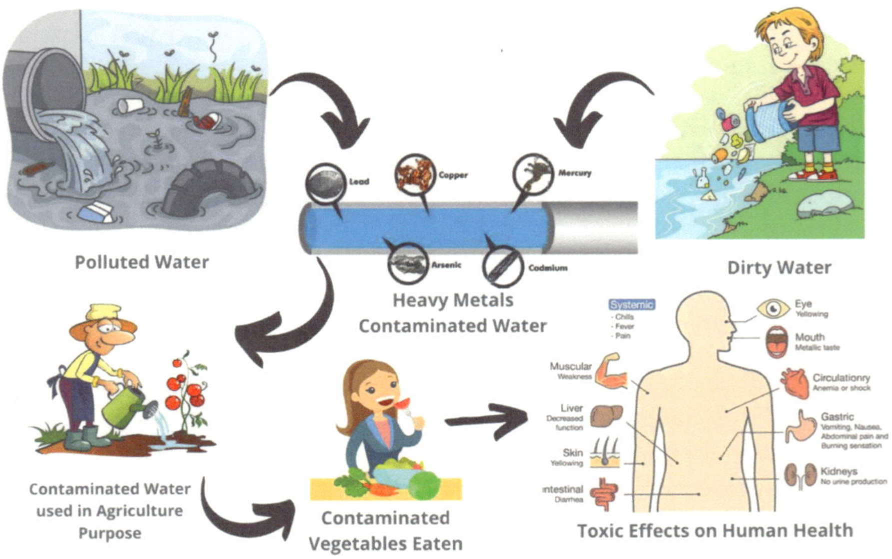

Figure 2.
Cyclic explanation of how vegetables contaminated and its toxic effects on humans.

DOI: http://dx.doi.org/10.5772/intechopen.102651

Free radicals are known to be produced by some heavy metals, which can cause oxidative stressing as well as other cellular damaging. The method by which free radicals are generated is unique to heavy metal. Heavy metals are acetified by the acid medium of stomach when they are consumed through food or drink. They oxidised to several oxidative states (Zn2+, Cd2+, Pb2+, As2+, As3+, Ag+, Hg2+, etc.) in this acidic media, which can quickly fix to biological molecules like proteins as well as enzymes to create persistent and strong connections. The thio groups are the most prevalent functional groups that heavy metals fixes to (SH group of cysteine and SCH3 group of methionine). Cadmium had shown to bind to cysteine remains in the catalytic surface of human thiol transfers in vitro, consisting thioredoxin reductase, glutathione reductase, as well as thioredoxin [65–70].

Heavy metal-bounded proteins might be able to be useful as a substratum by some enzymes. The heavy metal-bounded protein has an enzyme-substrate complex in a specific pattern, which prevents the enzyme through absorbing any more substrates till it is release. Resulting of the enzyme being inhibited, the product of substratum is not formed, and the heavy metal becomes embedded in the tissue, producing dysfunctions, abnormalities, and damage. Constraining thiol transferases reasons an rise in oxidative pressure and cell damaging. Poisonous arsenic, which can be there in fungicides, herbicides, and insecticides, can damage enzymes' –SH groups, preventing them from catalysing reactions.

As arsenite-inducing protein clustering was found and proved to be concentration-dependending, heavy metals may cause proteins to aggregate. The clusters also comprised a diverse ranging of proteins with roles linked to metabolism, protein portable, synthesis of protein, and protein stability [71–75]. After exposing to equi-toxic quantities of cadmium, arsenite, as well as chromium (Cr(VI)), Saccharomyces cerevisiae (budding yeast) cells gathered aggregated proteins, and the outcome of heavy metals on protein aggregation was altered in this direction: arsenic > cadmium > chromium [76–80]. The effectiveness of this agents' cellular uptake/export, as well as their different modalities of biological action, are likely to determine their in vivo potency to cause protein aggregation.

5. Heavy metal hazardous effects on human health

Heavy metals in soil, air, as well as water are a severe concerned since they will have a detrimental impact on food sustainability and human health. Eating of heavy metal-contaminating vegetables can result in a variety of ailments in consumers. Vegetable eating is the primary route for heavy metals to infect humans. Heavy metal pollution in food may produce heavy metal buildup in humans' kidneys and livers, disrupting a variety of biochemical processes that can lead to cardiovascular, neurological, renal, and bone illnesses [35, 81–84]. The biotoxic effects of high are determined by their concentrations and oxidation states, deposition mechanism, chemical composition of plants, physical characterisation, and rate of intake (**Table 2**) [1].

Cd had being discovered to have deleterious effects on a number of essential enzymes. The negative repercussions might include everything from a painful bone condition called ostemalacia to red blood cell disintegration and renal issues. High lead in the blood can induce hypertension, nephritis, and cardiovascular illness, as well as affecting children's cognitive development [61, 93, 94]. Cu as well as Zn can lead to acute stomach and bowel issues as well as liver damage [95–97]. Arsenic exposurance is linked to angiosarcoma and skin cancer [98, 99]. Zn, on

Heavy metal	Applications	Health effects	References
Chromium	paints pigment, fungicide, Pesticide	Cancer, nephritis and ulceration	[16]
Lead	Plastic, batteries, Auto exhaust, gasoline	Risk of cardiovascular disease and neurotoxic diseases	[85]
Cadmium	Pigments Fertiliser, plastic	Endocrine disrupter Carcinogenic, Alter calcium regulation in biological systems mutagenic, lung damage, fragile bones	[86, 87]
Zinc	Fertilisers	Dizziness, fatigue, vomiting, renal damage, decreased Immune function	[88, 89]
Nickle	Electroplating	Lung cancer, Immuno-toxic Allergic disease, neurotoxic, genotoxic, Infertility	[90]
Copper	Electronics, wood preservative, Architecture	Brain damage, Chronic anaemia, Kidney damage, Intestine irritation, Liver cirrhosis, Spontaneous abortions and gestational diabetes	[91]
Arsenic	Pesticides, Wood products & herbicides	Immunological, Reproductive and Developmental alterations and causing cancer	[92]
Mercury	Catalysts, Electric Switches, rectifiers, CFLs	Neurological and immune disorders, fatal to kidney and lungs	[27]

Table 2.
Various heavy metals, their application areas/industries, and probable harmful health consequences on humans produced by these heavy metals are shown.

the different side, can impair immunological function and raise stages of higher-density lipoproteins [99].

Due to higher heavy metal concentrations in the soil, fruit, as well as vegetables, the Vanregion of Turkey has a higher incidence of greater gastrointestinal cancer rates. Eating of heavymetal-contaminating food can depletes some vital bodily nutrients, resulting in lowered immune defences, altered physico-social behaviour, intrauterine growing retardation, and problems linked with malnourishment [100, 101]. Metal poisoning has also been linked to neurotoxic, carcinogenic, mutagenic, or teratogenic consequences, which might be acute, chronic, or sub-chronic. Some employees also stated having problems with their kidneys [102, 103].

The link between heavy metal exposure during pregnancy and foetal development has been widely established. Heavy metals have the potential to harm the reproductive system of female by causing damage to the ovary and hormone production and release [104, 105]. [106] found that heavy metals can causing alterations in the structure and role of the ovary, as well as embryonic development, when they were researched on the female reproductive system. In vivo and in vitro investigations have confirmed the deposing of heavy metals in the ovary. Pb in the body of the host has been linked to lower birth weightiness, preterm birthing, stillbirths, spontaneous abortions, as well as hypertension [107], while Ar in the body of the host has been linked to foetal loss, stillbirths, spontaneous abortions, and impaired growth as well as development [107, 108]. While Cd exposure is linked to low birth weight, AS exposure has been linked to spontaneous abortions and neurotoxic consequences. Cu poisoning is linked to lower birth weightiness, spontaneous abortions, and gestational diabetes [109]. [110] found women who had miscarriages had high methylmercury levels, albeit the link among methyl mercury exposure and spontaneous abortion has yet to be shown [110]. Stillbirths, miscarriages, and foetal development problems have described as a effect of mercury toxicity.

6. Future prospects and conclusion

Pollutants in the environment, food safety and security, and human health are all intricately intertwined. Heavy metal concentrations in the environment have risen rapidly in recent years. Heavy metal sources in vegetables differ across the developing and industrialised worlds. The principal contamination causes in soil–crop systems in industrialised nations are the deposition of PM on food plants and the usege of industrial effluents and sewage sludge as fertilisers. However, in underdeveloped nations, irrigation with untreated sewage or sludge is the primary cause of contamination for food crops. Heavy metal transmission from soil to crop systems is complicated and employs a variety of methods. To establish the true metal toxicity of multi-metal toxicity in vegetables, special care must be used. Human health hazards have been extensively investigated on a universal basis, but only a handful of these findings employed suitable epidemiological methodology. Existing control methods focus on decreasing heavy metal concentrations in soil and the food chain to decrease health hazards. To minimise the passage of metallic pollutants into the food chain and to develop appropriate remediation techniques, soil pollution must be mapped quickly and precisely. For temperately contaminating soils, biological remediation, such as phytoremediation and PGPR, could be a cost-effective and environmentally friendly alternative. With specific financial assurances, eco-friendly technical advancements such as nano-tools and farmer knowledge might benefit local economies and livelihoods.

Author details

Seema Manwani[1], Vanisree C.R.[2], Vibha Jaiman[1], Kumud Kant Awasthi[1], Chandra Shekhar Yadav[3], Mahipal Singh Sankhla[4], Pritam P. Pandit[4] and Garima Awasthi[1*]

1 Department of Life Sciences, Vivekananda Global University, Jaipur, India

2 Department of Forensic Sciences, SAGE University, Indore, India

3 School of Forensic Sciences, National Forensic Science University, Gandhinagar India

4 Department of Forensic Science, Vivekananda Global University, Jaipur, India

*Address all correspondence to: garima.awasthi@vgu.ac.in

References

[1] Duruibe JO, Ogwuegbu MOC, Egwurugwu JN. Heavy metal pollution and human biotoxic effects. International Journal of Physical Sciences. 2007;**2**(5):112-118

[2] Jabeen F, Aslam A, Salman M. Heavy metal contamination in vegetables and soil irrigated with sewage water and associated health risks assessment. Journal of Environmental and Agricultural Sciences. 2020;**22**(1):23-31

[3] Sankhla MS, Kumari M, Nandan M, Kumar R, Agrawal P. Heavy metals contamination in water and their hazardous effect on human health-a review. International Journal of Current Microbiology and Applied Sciences. 2016;**5**(10):759-766

[4] Hembrom S, Singh B, Gupta SK, Nema AK. A comprehensive evaluation of heavy metal contamination in foodstuff and associated human health risk: A global perspective. In: Contemporary Environmental Issues and Challenges in Era of Climate Change. Singapore: Springer; 2020. pp. 33-63

[5] Balkhair KS, Ashraf MAJSJOBS. Field accumulation risks of heavy metals in soil and vegetable crop irrigated with sewage water in western region of Saudi Arabia. Saudi Journal of Biological Sciences. 2016;**23**(1):S32-S44

[6] Sonone SS, Jadhav S, Sankhla MS, Kumar R. Water contamination by heavy metals and their toxic effect on aquaculture and human health through food chain. Letters in Applied NanoBioScience. 2020;**10**(2):2148-2166

[7] Boyd RS, Rajakaruna N. Heavy Metal Tolerance. Oxford, UK: Oxford University Press; 2013

[8] Christou A, Eliadou E, Michael C, Hapeshi E, Fatta-Kassinos D. Assessment of long-term wastewater irrigation impacts on the soil geochemical properties and the bioaccumulation of heavy metals to the agricultural products. Environmental monitoring and assessment. 2014;**186**(8):4857-4870

[9] Koivula MJ, Kanerva M, Salminen JP, Nikinmaa M, Eeva T. Metal pollution indirectly increases oxidative stress in great tit (Parus major) nestlings. Environmental Research. 2011;**111**(3):362-370

[10] Gupta N, Yadav KK, Kumar V, Krishnan S, Kumar S, Nejad ZD, et al. Evaluating heavy metals contamination in soil and vegetables in the region of North India: Levels, transfer and potential human health risk analysis. Environmental Toxicology and Pharmacology. 2021;**82**:103563

[11] Pollard AJ, Reeves RD, Baker AJ. Facultative hyperaccumulation of heavy metals and metalloids. Plant Science. 2014;**217**:8-17

[12] Proshad R, Kormoker T, Islam MS, Chandra K. Potential health risk of heavy metals via consumption of rice and vegetables grown in the industrial areas of Bangladesh. Human and Ecological Risk Assessment: An International Journal. 2019;**24**(4);921-943

[13] Kharazi A, Leili M, Khazaei M, Alikhani MY, Shokoohi R. Human health risk assessment of heavy metals in agricultural soil and food crops in Hamadan, Iran. Journal of Food Composition and Analysis. 2021;**100**:103890

[14] Clemens S, Ma JF. Toxic heavy metal and metalloid accumulation in crop plants and foods. Annual Review of Plant Biology. 2016;**67**:489-512

[15] Chen Z, Muhammad I, Zhang Y, Hu W, Lu Q, Wang W, et al. Transfer of heavy metals in fruits and vegetables grown in greenhouse cultivation systems and their health risks in Northwest China. Science of the Total Environment. 2021;**766**:142663

[16] Onakpa MM, Njan AA, Kalu OC. A review of heavy metal contamination of food crops in Nigeria. Annals of Global Health. 2018;**84**(3):488

[17] Khan MA, Majeed R, Fatima SU, Khan MA, Shahid S. Occurrence, distribution and health effects of heavy metals in commercially available vegetables in Karachi. International Journal of Biology and Biotechnology. 2020;**17**:319-328

[18] Oguh CE, Obiwulu ENO. Human risk on heavy metal pollution and bioaccumulation factor in soil and some edible vegetables around active auto-mechanic workshop in Chanchaga Minna Niger state, Nigeria. Annals of Ecology and Environmental Science. 2020;**4**(1):12-22

[19] Awasthi G, Nagar V, Mandzhieva S, Minkina T, Sankhla MS, Pandit PP, et al. Sustainable amelioration of heavy metals in soil ecosystem: Existing developments to emerging trends. Minerals. 2022;**12**(1):85

[20] Ogwuegbu MO, Ijioma MA. Effects of certain heavy metals on the population due to mineral exploitation. In: International Conference on Scientific and Environmental Issues In the Population, Environment and Sustainable Development in Nigeria. Ekiti State, Nigerian: University of Ado Ekiti; 2003. pp. 8-10

[21] Sridhar BM, Vincent RK, Roberts SJ, Czajkowski K. Remote sensing of soybean stress as an indicator of chemical concentration of biosolid amended surface soils. International Journal of Applied Earth Observation and Geoinformation. 2011;**13**(4):676-681

[22] Liu X, Gu S, Yang S, Deng J, Xu J, et al. Heavy metals in soil-vegetable system around E-waste site and the health risk assessment. Science of The Total Environment. 2021;**779**:146438

[23] Vatanpour N, Feizy J, Talouki HH, Es'haghi Z, Scesi L, Malvandi AM. The high levels of heavy metal accumulation in cultivated rice from the Tajan river basin: Health and ecological risk assessment. Chemosphere. 2020;**245**:125639

[24] Su Y, Sridhar, MBB, Han FX, Diehl SV, Monts DL. Effect of bioaccumulation of Cs and Sr natural isotopes on foliar structure and plant spectral reflectance of Indian mustard (Brassica juncea). Water, Air, and Soil Pollution. 2007;**180**(1):65-74

[25] Sridhar MB, Han FX, Diehl SV, Monts DL, Su Y. Monitoring the effects of arsenic and chromium accumulation in Chinese brake fern (Pteris vittata). International Journal of Remote Sensing. 2007;**28**(5):1055-1067

[26] Dixit G, Singh AP, Kumar A, Mishra S, Dwivedi S, Kumar S, et al. Reduced arsenic accumulation in rice (Oryza sativa L.) shoot involves sulfur mediated improved thiol metabolism, antioxidant system and altered arsenic transporters. Plant Physiology and Biochemistry. 2016;**99**:86-96

[27] Manzoor J, Sharma M, Wani KA. Heavy metals in vegetables and their impact on the nutrient quality of vegetables: A review. Journal of plant Nutrition. 2018;**41**(13): 1744-1763

[28] Dixit G, Singh AP, Kumar A, Singh PK, Kumar S, Dwivedi S, et al. Sulfur mediated reduction of arsenic toxicity involves efficient thiol metabolism and the antioxidant defense system in rice. Journal of Hazardous Materials. 2015;**298**:241-251

[29] Mohod CV. A review on the concentration of the heavy metals in vegetable samples like spinach and tomato grown near the area of Amba Nalla of Amravati City. International Journal of Innovative Research in Science, Engineering and Technology. 2015;**4**(5):2788-2792

[30] Zhou H, Yang WT, Zhou X, Liu L, Gu JF, Wang WL, et al. Accumulation of heavy metals in vegetable species planted in contaminated soils and the health risk assessment. International Journal of Environmental Research and Public Health. 2016;**13**(3):289

[31] Islam MS, Hoque MF. Concentrations of heavy metals in vegetables around the industrial area of Dhaka city, Bangladesh and health risk assessment. International Food Research Journal. 2014;**21**(6):2121

[32] Shakya PR, Khwaounjoo NM. Heavy metal contamination in green leafy vegetables collected from different market sites of Kathmandu and their associated health risks. Scientific World. 2013;**11**(11):37-42

[33] Kachenko A, Singh B. Heavy metals contamination of home grown vegetables near metal smelters in NSW. In: 3rd Australian New Zealand Soils Conference. Australia: The regional institute online publishing; 2004. pp. 5-9

[34] Chauhan G. Toxicity study of metals contamination on vegetables grown in the vicinity of cement factory. International Journal of Scientific and Research Publication. 2014;**4**(11):1-8

[35] Sharma RK, Agrawal M, Marshall FM. Heavy metals in vegetables collected from production and market sites of a tropical urban area of India. Food and Chemical Toxicology. 2009;**47**(3):583-591

[36] López-Millán AF, Sagardoy R, Solanas M, Abadía A, Abadía J. Cadmium toxicity in tomato (Lycopersicon esculentum) plants grown in hydroponics. Environmental and Experimental Botany. 2009;**65**(2-3):376-385

[37] Singh AP, Dixit G, Kumar A, Mishra S, Singh PK, Dwivedi S, et al. Nitric oxide alleviated arsenic toxicity by modulation of antioxidants and thiol metabolism in rice (Oryza sativa L.). 2016;**6**:1272

[38] Chen Y, He YF, Luo YM, Yu YL, Lin Q, Wong MH. Physiological mechanism of plant roots exposed to cadmium. Chemosphere. 2003;**50**(6):789-793

[39] Awasthi G, Singh T, Awasthi A, Awasthi KK. Arsenic in mushrooms, fish, and animal products. In: Arsenic in Drinking Water and Food. Singapore: Springer; 2020. pp. 307-323

[40] Lin R, Wang X, Luo Y, Du W, Guo H, Yin D. Effects of soil cadmium on growth, oxidative stress and antioxidant system in wheat seedlings (Triticum aestivum L.). Chemosphere. 2007;**69**(1):89-98

[41] Singh AP, Dixit G, Kumar A, Mishra S, Kumar N, Dixit S, et al. A protective role for nitric oxide and salicylic acid for arsenite phytotoxicity in rice (Oryza sativa L.). Plant Physiology and Biochemistry. 2017;**115**:163-173

[42] McLaughlin MJ, Tiller KG, Naidu R, Stevens DP. The behaviour and

environmental impact of contaminants in fertilizers. Soil Research. 1996;**34**(1):1-54

[43] Dave R, Tripathi RD, Dwivedi S, Tripathi P, Dixit G, Sharma YK, et al. Arsenate and arsenite exposure modulate antioxidants and amino acids in contrasting arsenic accumulating rice (Oryza sativa L.) genotypes. Journal of Hazardous Materials. 2013;**262**:1123-1131

[44] Dixit G, Singh AP, Kumar A, Dwivedi S, Deeba F, Kumar S, et al. Sulfur alleviates arsenic toxicity by reducing its accumulation and modulating proteome, amino acids and thiol metabolism in rice leaves. Scientific Reports. 2015;**5**(1):1-16

[45] Koo BJ, Chang AC, Crowley DE, Page AL, Taylor A. Availability and plant uptake of biosolid-borne metals. Applied and Environmental Soil Science. 2013;**2013**

[46] Dave R, Singh PK, Tripathi P, Shri M, Dixit G, Dwivedi S, et al. Arsenite tolerance is related to proportional thiolic metabolite synthesis in rice (Oryza sativa L.). Archives of Environmental Contamination and Toxicology. 2013;**64**(2):235-242

[47] Xiong ZT, Wang H. Copper toxicity and bioaccumulation in Chinese cabbage (Brassica pekinensis Rupr.). Environmental Toxicology: An International Journal. 2005;**20**(2):188-194

[48] Yang XE, Long XX, Ni WZ, Ye ZQ, He ZL, Stoffella PJ, et al. Assessing copper thresholds for phytotoxicity and potential dietary toxicity in selected vegetable crops. Journal of Environmental Science and Health, Part B. 2002;**37**(6):625-635

[49] Kumar A, Dixit G, Singh AP, Dwivedi S, Srivastava S, Mishra K, Tripathi RD. Selenate mitigates arsenite toxicity in rice (Oryza sativa L.) by reducing arsenic uptake and ameliorates amino acid content and thiol metabolism. Ecotoxicology and Environmental Safety. 2016;**133**:350-359

[50] John R, Ahmad P, Gadgil K, Sharma S. Heavy metal toxicity: Effect on plant growth, biochemical parameters and metal accumulation by Brassica juncea L. 2009;**3**(3):66-75

[51] Xu J, Yang L, Wang Z, Dong G, Huang J, Wang Y. Toxicity of copper on rice growth and accumulation of copper in rice grain in copper contaminated soil. Chemosphere. 2006;**62**(4):602-607

[52] Lokhande VH, Patade VY, Srivastava S, Suprasanna P, Shrivastava M, Awasthi G. Copper accumulation and biochemical responses of Sesuvium portulacastrum (L.). Materials Today: Proceedings. 2020;**31**:679-684

[53] Abou Auda M, Ali EES. Cadmium and zinc toxicity effects on growth and mineral nutrients of carrot (Daucus carota). Pakistan Journal of Botany. 2010;**42**:341-351

[54] Zengin FK, Munzuroglu O. Toxic effects of cadmium (Cd++) on metabolism of sunflower (Helianthus annuus L.) seedlings. Acta Agriculturae Scandinavica Section B-Soil and Plant Science. 2006;**56**(3):224-229

[55] Krishna AK, Mohan KR. Distribution, correlation, ecological and health risk assessment of heavy metal contamination in surface soils around an industrial area, Hyderabad, India. Environmental Earth Sciences. 2016;**75**(5):1-17

[56] Fonge BA, Larissa MT, Egbe AM, Afanga YA, Fru NG, Ngole-Jeme VM. An

assessment of heavy metal exposure risk associated with consumption of cabbage and carrot grown in a tropical Savannah region. Sustainable Environment. 2021;**7**(1):1909860

[57] Feseha A, Chaubey AK, Abraha A. Heavy metal concentration in vegetables and their potential risk for human health. Health Risk Analysis. 2021;**1**:68-81

[58] Sankhla MS, Kumari M, Sharma K, Kushwah RS, Kumar R. Heavy metal pollution of Holy River ganga: A review. International Journal of Research. 2018;**5**(1):421-436

[59] Yadav H, Kumar R, Sankhla MS. Residues of pesticides and heavy metals in crops resulting in toxic effects on living organism. Journal Seybold Report. 2020;**1533**:9211

[60] Florea AM, Büsselberg D. Occurrence, use and potential toxic effects of metals and metal compounds. Biometals. 2006;**19**(4):419-427

[61] Sankhla MS, Kumar R, Prasad L. Zinc impurity in drinking water and its toxic effect on human health. Indian Congress of Forensic Medicine & Toxicology. 2019;**17**(4):84-87

[62] Yu M-H, Tsunoda H. Environmental Toxicology: Biological and Health Effects of Pollutants. Boca Raton, Florida: CRC Press; 2004

[63] Sangameshwar R, Rasool A, Venkateshwar C. Effect of heavy metals on leafy vegetable (trigonella foenum-graecum l.) and its remediation. Plant Archives. 2020;**20**(2):1941-1944

[64] Sulaiman FR, Ibrahim NH, Ismail SN. Heavy metal (As, Cd, and Pb) concentration in selected leafy vegetables from Jengka, Malaysia, and potential health risks. SN Applied Sciences. 2020;**2**(8):1-9

[65] Chrestensen CA, Starke DW, Mieyal JJ. Acute cadmium exposure inactivates thioltransferase (Glutaredoxin), inhibits intracellular reduction of protein-glutathionyl-mixed disulfides, and initiates apoptosis. Journal of Biological Chemistry. 2000;**275**(34):26556-26565

[66] Haque MM, Niloy NM, Khirul MA, Alam MF, Tareq SM. Appraisal of probabilistic human health risks of heavy metals in vegetables from industrial, non-industrial and arsenic contaminated areas of Bangladesh. Heliyon. 2021;**7**(2): e06309

[67] Ogunwale T, Ogar P, Kayode G, Salami K, Oyekunle J, Ogunfowokan A. Health risk assessment of heavy metal toxicity utilizing eatable vegetables from poultry farm region of Osun state. Journal of Environment Pollution and Human Health. 2021;**9**(1):6-15

[68] Can H, Ozyigit II, Can M, Hocaoglu-Ozyigit A, Yalcin IE. Environment-based impairment in mineral nutrient status and heavy metal contents of commonly consumed leafy vegetables marketed in Kyrgyzstan: A case study for health risk assessment. Biological Trace Element Research. 2021;**199**(3):1123-1144

[69] Sankhla MS, Sharma K, Kumar R. Heavy metal causing neurotoxicity in human health. International Journal of Innovative Research in Science. Engineering and Technology. 2017;**6**:5

[70] Sankhla M, Kumari M, Nandan M, Kumar R, Agrawal P. Heavy metal contamination in soil and their toxic effect on human health: A review study. International journal of All Research Education and Scientific Methods. 2016;**4**:13-19

[71] Tamás MJ, Sharma SK, Ibstedt S, Jacobson T, Christen P. Heavy metals and metalloids as a cause for protein misfolding and aggregation. Biomolecules. 2014;**4**(1):252-267

[72] Othman YA, Al-Assaf A, Tadros MJ, Albalawneh A. Heavy metals and microbes accumulation in soil and food crops irrigated with wastewater and the potential human health risk: A metadata analysis. Water. 2021;**13**(23):3405

[73] Gebeyehu HR, Bayissa LD. Levels of heavy metals in soil and vegetables and associated health risks in mojo area, Ethiopia. Plos One. 2020;**15**(1):e0227883

[74] Rani J, Agarwal T, Chaudhary S. Health risk assessment of heavy metals through the consumption of vegetables in the National Capital Region, India. 2021

[75] Sankhla MS, Kumar R, Prasad L. Estimation of zinc concentration in Yamuna River (Delhi) water due to climatic changes. Journal of Punjab Academy of Forensic Medicine & Toxicology. 2021;**21**(1)

[76] Jacobson T, Navarrete C, Sharma SK, Sideri TC, Ibstedt S, Priya S, et al. Arsenite interferes with protein folding and triggers formation of protein aggregates in yeast. Journal of Cell Science 2012;**125**(21):5073-5083

[77] Nambafu GN. Extent of heavy metals contamination in leafy vegetables among Peri-urban farmers. Asian Journal of Research in Botany. 2020:38-46

[78] Sankhla MS, Kumar R, Biswas A. Dynamic nature of heavy metal toxicity in water and sediments of Ayad River with climatic change. International Journal of Hydrogen Energy. 2019;**3**(5):339-343

[79] Parihar K, Kumar R, Sankhla MS. Impact of heavy metals on survivability of earthworms. International Medico-Legal Reporter Journal. 2019;**2**

[80] Verma RK, Sankhla MS, Jadhav EB, Parihar K, Awasthi KK. Phytoremediation of heavy metals extracted soil and aquatic environments: Current advances as well as emerging trends. Biointerface Research in Applied Chemistry. 2021;**12**:5486-5509

[81] Sankhla MS, Kumar R. Contaminant of heavy metals in groundwater & its toxic effects on human health & environment. 2019;**3**(7);945-949

[82] Verma RK, Sankhla MS, Kumar R. Mercury contamination in Water & its Impact on public health. International Journal of Forensic Science. 2018;**1**(2)

[83] Pateriya A, Verma RK, Sankhla MS, Kumar R. Heavy metal toxicity in rice and its effects on human health. Letters in Applied NanoBioScience. 2020;**10**(1):1833-1845

[84] Jiang HH, Cai LM, Wen HH, Hu GC, Chen LG, Luo, J. An integrated approach to quantifying ecological and human health risks from different sources of soil heavy metals. Science of the Total Environment. 2020;**701**:134466

[85] Kumar A, Chaturvedi AK, Yadav K, Arunkumar KP, Malyan SK, Raja P. et al. Fungal phytoremediation of heavy metal-contaminated resources: Current scenario and future prospects. In: Recent Advancement in White Biotechnology through Fungi. Cham, Switzerland: Springer; 2019. pp. 437-461

[86] Sharma T, Banerjee BD, Yadav CS, Gupta P, Sharma S. Heavy metal levels in adolescent and maternal blood: Association with risk of hypospadias.

International Scholarly Research Notices. 2014;**2014**

[87] He Z, Shentu J, Yang X, Baligar VC, Zhang T, Stoffella PJ. Heavy metal contamination of soils: Sources, indicators and assessment. 2015;**9**:17-18

[88] Mishra S, Bharagava RN, More N, Yadav A, Zainith S, Mani S, Chowdhary P, et al. Heavy metal contamination: An alarming threat to environment and human health. In: Environmental Biotechnology: For Sustainable Future. Singapore: Springer; 2019. pp. 103-125

[89] Benvenuti T, Krapf RS, Rodrigues MAS, Bernardes AM, Zoppas-Ferreira J. Recovery of nickel and water from nickel electroplating wastewater by electrodialysis. Separation and Purification Technology. 2014;**129**:106-112

[90] Wang YP, Shi JY, Qi LIN, Chen XC, Chen YX. Heavy metal availability and impact on activity of soil microorganisms along a Cu/Zn contamination gradient. Journal of Environmental Sciences. 2007;**19**(7):848-853

[91] Henriques B, Lopes CB, Figueira P, Rocha LS, Duarte AC, Vale C, et al. Bioaccumulation of Hg, Cd and Pb by Fucus vesiculosus in single and multi-metal contamination scenarios and its effect on growth rate. Chemosphere. 2017;**171**:208-222

[92] Skalnaya MG, Tinkov AA, Lobanova YN, Chang JS, Skalny AV. Serum levels of copper, iron, and manganese in women with pregnancy, miscarriage, and primary infertility. Journal of Trace Elements in Medicine and Biology 2019;**56**:124-130

[93] Ekong EB, Jaar BG, Weaver VM. Lead-related nephrotocixity: A review of the epidemiologic evidence. Kidney Int. 2006;**70**:2074-2084

[94] Navas-Acien A, Guallar E, Silbergeld EK, Rothenberg SJ. Lead exposure and cardiovascular disease—a systematic review. Environmental Health Perspectives. 2007;**115**:472-482

[95] Hartley W, Lepp NW. Remediation of arsenic contaminated soils by iron-oxide application, evaluated in terms of plant productivity, arsenic and phytotoxic metal uptake. Science of the Total Environment. 2008;**390**(1): 35-44

[96] Sankhla MS, Kumar R, Agrawal P, et al. Arsenic in water contamination & toxic effect on human health: Current scenario of India. Journal of Forensic Sciences & Criminal Investigation. 2018;**10**(2):001-5

[97] Ali H, Khan E. Bioaccumulation of non-essential hazardous heavy metals and metalloids in freshwater fish. Risk to Human Health. Environmental Chemistry Letters. 2018;**16**(3):903-917

[98] Rahman MA, Rahman MM, Reichman SM, Lim RP, Naidu R. et al. Heavy metals in Australian grown and imported rice and vegetables on sale in Australia: Health hazard. Ecotoxicology and Environmental Safety. 2014;**100**:53-60

[99] Harmanescu M, Alda LM, Bordean DM, Gogoasa I, Gergen I, et al. Heavy metals health risk assessment for population via consumption of vegetables grown in old mining area; a case study: Banat County, Romania. Chemistry Central Journal. 2011;**5**(1):1-10

[100] Mapanda F, Mangwayana EN, Nyamangara J, Giller KE, et al. The effect of long-term irrigation using wastewater on heavy metal contents of soils under vegetables in Harare, Zimbabwe. Agriculture, Ecosystems & Environment. 2005;**107**(2-3):151-165

[101] Kohzadi S, Shahmoradi B, Ghaderi E, Loqmani H, Maleki A, et al. Concentration, source, and potential human health risk of heavy metals in the commonly consumed medicinal plants. Biological Trace Element Research. 2019;**187**(1):41-50

[102] Kadir MM, Janjua NZ, Kristensen S, Fatmi Z, Sathiakumar N. Status of children's blood lead levels in Pakistan: Implications for research and policy. Public Health. 2008;**122**(7):708-715

[103] Rai PK, Lee SS, Zhang M, Tsang YF, Kim KH. Heavy metals in food crops: Health risks, fate, mechanisms, and management. Environment International. 2019;**125**:365-385

[104] Silbergeld EK. Lead in bone: Implications for toxicology during pregnancy and lactation. Environmental Health Perspectives. 1991;**91**:63-70

[105] Sankhla MS, Kumar R, and Prasad L. Distribution and contamination assessment of potentially harmful element chromium in water. 2019;**2**(3). Available at SSRN 3492307

[106] Kumar S. Occupational exposure associated with reproductive dysfunction. Journal of Occupational Health. 2004;**46**(1):1-19

[107] Grant K, Goldizen FC, Sly PD, Brune MN, Neira M, van den Berg M, et al. Health consequences of exposure to e-waste: A systematic review. The Lancet Global Health. 2013;**1**(6):350-361

[108] Adimalla N, Wang H. Distribution, contamination, and health risk assessment of heavy metals in surface soils from northern Telangana, India. Arabian Journal of Geosciences. 2018;**11**(21):1-15

[109] Basu A, Yu Y, Jenkins AJ, Nankervis A, Hanssen K, Scholz H, et al. Plasma trace elements and preeclampsia in type 1 diabetes: A prospective study. In: Diabetes. Vol. 63. USA: American Diabetes Association; 2014. pp. A385-A385

[110] Benefice E, Luna-Monrroy S, Lopez-Rodriguez R, et al. Fishing activity, health characteristics and mercury exposure of Amerindian women living alongside the Beni River (Amazonian Bolivia). International Journal of Hygiene and Environmental Health. 2010;**213**(6):458-464